西安电子科技大学教材建设基金资助项目

科技汉英笔译

主　编　仝文宁

副主编　李长安　张　明

参　编　周正履　任利华　朱琳菲　罗　铮

主　审　秦荻辉

西安电子科技大学出版社

内 容 简 介

　　本书以培养译者的素养和家国情怀为宗旨，以提升译者的翻译能力和综合能力为目标，在内容编排上涉及科技翻译伦理、标准等诸多方面，并从词法、句法、篇章三个维度展示科技文本的翻译方法与注意事项。鉴于计算机技术在科技翻译领域中的广泛应用，如何使用机器翻译来提升翻译水平，也是本书的另一项重要内容。此外，本书还对翻译腔等常见问题进行分析，探索此类问题之根源及解决方案。本书在体例上始于权威案例赏析，并辅以大量案例解析来提高学习者的译文鉴赏能力及实际问题解决能力。

　　本书可作为高等院校英语类专业及翻译专业学位硕士的科技笔译课程教材，也可供有志于促进国际科技交流的科技工作者使用。

图书在版编目(CIP)数据

科技汉英笔译 / 仝文宁主编. --西安：西安电子科技大学出版社，2024.2
ISBN 978 - 7 - 5606 - 7182 - 6

Ⅰ.①科…　　Ⅱ.①仝…　　Ⅲ.①科学技术—英语—翻译—教材　　Ⅳ.①G301

中国国家版本馆 CIP 数据核字(2024)第 036714 号

策　　划　黄薇谚
责任编辑　高　樱　黄薇谚
出版发行　西安电子科技大学出版社（西安市太白南路 2 号）
电　　话　(029)88202421　88201467　　　　邮　　编　710071
网　　址　www.xduph.com　　　　　　　　　电子邮箱　xdupfxb001@163.com
经　　销　新华书店
印刷单位　咸阳华盛印务有限责任公司
版　　次　2024 年 2 月第 1 版　　　　　　　2024 年 2 月第 1 次印刷
开　　本　787 毫米×960 毫米　　　　　　　1/16　印张　15.5
字　　数　270 千字
定　　价　42.00 元
ISBN 978 - 7 - 5606 - 7182 - 6 / G
XDUP 7484001-1

＊＊＊ 如有印装问题可调换 ＊＊＊

前　言

随着人类命运共同体的构建和"一带一路"倡议的推进，我国与世界各国的科技交流日益深化。"十三五"以来，我国技术贸易规模持续扩大，技术贸易合作伙伴遍及世界各地，技术出口能力也不断增强。我国已从技术引进大国，成为重要的技术输出国。技术输出需要高水平、专业化的科技翻译人才。《普通高等学校本科翻译专业教学指南》也指出，翻译专业要培养能适应国家与地方经济建设和社会发展需要的复合型人才。本书正是基于国家科技发展现状和翻译专业人才培养使命，在科技文本汉译英方面所做的尝试和探索。

本书兼顾宏观翻译理论、原则和微观科技翻译实务两个层面，既涵盖科技翻译伦理、过程、方法，也从词法、句法、篇章三个维度讨论科技汉英翻译注意事项。同时，本书也关注译者，在书中探索翻译腔的本质和根源，并给出应对策略，包括学习方法、习惯和意识养成，译者精神培养等。具体来说，本书具有以下特色：

第一，注重价值观引领。本书较为系统地介绍了翻译伦理，使学习者了解中外翻译家的译论和情怀，增强译者的专业归属感和使命感。本书还对科技译者的职业道德准则进行了详细解读，以培养学习者的责任意识和职业素养。实践类章节设有"经典赏析"模块，精选《中国日报》"每日一词"中的科技相关话题，以汉英双语介绍当下国家大事及热门话题，紧扣时代主题，译文权威性强，内容涵盖政治、经济、文化、民生、科技等。通过赏析"每日一词"经典翻译，学习者在学习翻译技巧的同时，也可增强文化自信和家国情怀。

第二，强调综合能力培养。本书不再赘述翻译知识和技巧，转而关注翻译本身以及翻译的主体——译者，着力介绍翻译过程、标准、原则等，分析译者所面临问题的根源，提供解决问题的方案，以此培养学习者的思辨能力、终身学习能力、信息技术应用能力、实践能力和调查研究能力等。

第三，原创性强。本书所有案例解析均为编者亲自撰写。除少数经典例句来自《中国日报》外，本书大部分案例均来自编者的教学或翻译实践过程。书中有关原则、策略均基于广泛调研，可保证权威性，其中更多是编者长年累月观察、总结、反思、分析的成果，具有极强的原创性。

第四，跨学科性。翻译本身具有跨学科属性，本书则更加凸显科技性。首先，本书案例涉及电子、信息、云计算、新能源等诸多科技领域，为译者未来进入专业领域热身。其次，本书多个章节讨论译者的科技素养以及如何提高科技素养，从观点上支撑了本书的科技属性。最后，本书的所有编者均具有理工知识背景，且长期从事科技笔译工作，在专业上保证了本书的科技属性。

本书由西安电子科技大学外国语学院仝文宁担任主编，秦荻辉担任主审，仝文宁、李长安、张明、周正履、任利华、朱琳菲、罗铮合作完成编写工作。仝文宁负责第二章、第十一章、第十二章的编写及全书的审校工作，李长安负责第五章的编写及部分审校工作，张明负责第三章、第七章的编写及全书的统稿工作及部分审校工作，周正履负责第八章、第九章的编写及部分审校工作，任利华负责第四章、第十章的编写，朱琳菲负责第六章的编写，罗铮负责第一章的编写。

本书可作为高等院校英语类专业及翻译专业学位硕士的科技笔译课程教材，也可供有志于促进国际科技交流的科技工作者使用。

本书是西安电子科技大学教材建设基金资助项目，在编辑出版过程中得到了西安电子科技大学出版社的大力支持，在此一并表示感谢。同时感谢《中国日报》"每日一词"模块提供的经典翻译，感谢所有参考文献作者的真知灼见。

本书也是陕西省哲学社会科学研究项目"以国际传播为导向的翻译教材建设研究"(2023HZ0994)和教育部产学合作协同育人项目"新文科背景下理工科高校科技翻译师资及人才培养实践体系构建"(220900630181551)的部分成果。

本书观点、解析或可粗浅，但皆出于真诚，愿对学习者有所启发，若能引起共鸣，则更为欣慰。若有不足之处，祈请赐正。

编　者
2024 年 1 月于
西安电子科技大学外国语学院

目　录

第一章　翻　译　概　述

　　一般情况下，翻译是指将一种语言文字所传达的信息用另一种语言文字表达出来的过程。从这个定义出发，翻译常常被视为不同语言文字之间的转换。这虽然说出了翻译的本质，但翻译所蕴含的更为丰富的内涵却未被提及。实际上，从语言文字转换的角度认识翻译，只道出了翻译的过程，并未涉及翻译的结果、功能等内容。单从翻译的功能来看，翻译往往肩负着文化变革与革新的功能。中国历史上曾出现过三次翻译高潮：东汉至唐宋的佛经翻译、明末清初的科技翻译、清末民初的西学翻译。伴随每次翻译浪潮而来的，是大量外来思想的涌入以及对本国文化的革新。鉴于翻译对不同文明的重要贡献，鲁迅先生曾称翻译为"人类文明之火"，称译者为"盗火的普罗米修斯"。

　　单就翻译策略与翻译方法而言，国内的相关认识也在不断发生变化。长期以来，严复提出的"信、达、雅"一直被国内视为翻译的圭臬。改革开放以来，中外文化交流互通，翻译活动也日益频繁，人们对翻译活动的认识亦在不断深入。尤其是近30年来，西方翻译理论大规模涌入，国内对翻译标准和方法的探讨也不再仅仅停留于"信、达、雅"、直译、意译等方面，对动态对等、功能对等、语义翻译、交际翻译、翻译目的论等一系列西方翻译理论和方法的研究不断丰富。这些研究在推动国内翻译研究蓬勃发展的同时，也让不少初学者对这些外来的翻译理论和方法"一头雾水"，对翻译望而却步。要充分理解这些外来概念，就需要对其产生的背景和拟解决的问题进行全面深入的分析。

　　简单来说，翻译活动至少涉及"原文""译者""译文""读者"四个方面。译者将源语中所表达的内容经过文字转换用另一种语言表达出来，直接或经出版送到目标读者手中，读者通过阅读译文来了解原文所要表达的内容。而对译者、文本、读者的不同侧重，引发对翻译活动关注点的不同，导致各翻译流派和流派研究的重心不同。长期以来，国内对翻译活动的关注以文本为中心，即关注原文向译文转换。不管是早期针对佛经翻译道安和玄奘所提出的"五失本""三不易"和"五不翻"，还是严复提出的"信、达、雅"，傅雷提出的"神似"和钱钟书的"化境"，我国传统译论的很多探讨，本质上都是对翻译过程中文本转换标准、规范的设定与探讨。

从译者入手的翻译探究往往涉及翻译过程中译者所要遵守的翻译原则、翻译规范，或是对译者翻译过程中主体性的探究。近年来，大部分翻译伦理探究也是聚焦于译者所进行的考察。此外，随着翻译技术的蓬勃发展，利用眼动仪等高科技设备所做的翻译研究也往往以译者为着眼点，考察翻译过程中译者的认知与心理。而从读者角度入手的翻译研究，往往会关注翻译过程中针对目标语读者情况而对原文的改写和译文在目标语读者中的接受度。例如，朱生豪先生在翻译莎士比亚作品的过程中，考虑到我国传统文化和 20 世纪初普通民众的接受度，就将莎士比亚作品中露骨的情色描写统统删除，向中国读者呈现出和原文略有差异的莎剧译文。

此外，翻译也不可能在真空中进行，译文和原文的外部政治、经济、文化因素也往往会对文本造成很大影响。比如，中苏关系就曾对苏俄文学的译入产生重大影响。在新中国成立不久，我国翻译了大量苏俄文学作品。据统计，自1949 年至 1958 年的 10 年间，中国翻译苏俄文学作品达 3526 种，占全部外国文学翻译总数和印数的四分之三；然而，随着 20 世纪 60 年代中苏关系冷却，苏俄文学译入数量也逐年递减(赵稀方，2009)。一些西方汉学家译者在翻译当代中国文学的过程中，也往往喜欢选译那些反映新中国经济发展阵痛的文本，以凸显中国社会发展过程中的矛盾与冲突，加剧西方社会对我国的误解与偏见。因此，这就要求在翻译过程中，译者"具备良好的政治思想素质，保持鲜明的政治立场，具有高度的思想政治觉悟和政治敏感度"(黄友义，2023)。

经济因素对翻译的影响，主要体现在翻译报酬、翻译出版等方面。文化因素对翻译的影响，主要体现为中外文化差异对翻译所造成的影响。由于文化的概念本身就有广义和狭义之分，因而文化对翻译的影响小到词汇、短语、句法层面的调整，大到篇章结构、诗学范式，可谓无所不包。

可以看出，翻译并非仅仅是两种语言文字间的转换，而是涉及更广泛的内容，但对于初涉翻译的译者，要做的第一步是先掌握翻译的一般策略与方法，在翻译实践中灵活运用翻译策略与方法，为读者提供高水平的译文。本书聚焦科技文本翻译的探讨，鉴于科技文本不同于文学文本、法律文本的独特性，因而本书会针对科技性文本自身的语言特点，介绍科技文本翻译的注意事项及翻译过程中常用的策略与方法，并结合机器翻译对科技文本翻译的辅助作用进行讲解，从而使读者对科技文本翻译具备较全面客观的了解，更好地掌握科技翻译。

第一节 科技翻译与功能翻译理论

科技翻译实质上是对科技文本的翻译，之所以将科技文本翻译单独探讨，主要原因是科技文本具有不同于文学文本、法律文本的语言特征。相较于文学文本，科技文本总体上具有内容的专业性、表达的简洁性等特征。鉴于科技文本的独特语言特征，在翻译过程中往往需要采用更具针对性的策略与方法，从而达到更佳的翻译效果。科技文本翻译在更深层次上体现了对所译文本按类划分，并基于不同类型文本特点进行翻译的思想。为什么要明确指出这一点呢？因为在较长一段时期对翻译标准的制定不是依据文本类型进行区分的，人们往往会把同一个翻译标准应用于不同类型的文本翻译之中。这就给刚入门的译者一种错觉，好像文学文本的翻译方法与标准和科技、法律等其他类型文本并无不同。实际上，文本类型不同，翻译过程中往往也会表现出不同特点。因此，翻译过程中译者应根据不同文本类型的语言特点，因地制宜地选择更适合该文本语言特点的翻译方法，从而提供更高水平的译文。

传统上对文体的划分主要是依据写作手法，将文章分为"议论文、说明文、记叙文、抒情文等"(袁晖、李熙宗，2005)。刘宓庆(2012)在《文体与翻译》中主要探讨了六种文体的翻译，即新闻报刊文体、论述文体、公文文体、描写及叙述文体、科技文体及应用文体，然而针对这六种文体的分类标准与依据刘宓庆并未作出详细说明。李长栓(2012)将文本类型直接划分为"文学文本"和"非文学文本"，并指出两种类型文本的区别主要体现在"内容""形式""思维方式""读者预期""文本功能"五大方面。

西方学者基于文本类型划分对翻译的探讨首推德国学者凯瑟琳娜·莱斯(Katharina Reiss)。莱斯(2004)认为译者与批判家对翻译质量的评估应该基于相同的依据——语言，而文本需要以语言为媒介进行表达，因此需要对每个文本意欲再现的功能进行考察。继而，莱斯(2004)基于卡尔·布勒(Karl Buhler)对语言的再现、表达、劝说功能的划分，认为这三种语言功能分别强调了语言的逻辑、审美和对话维度，并与之相对形成信息型文本、表达型文本和操作型文本[1]

---

[1] 关于文本类型的称谓，不同著作的译文略有差异。莱斯语篇划分图中的三个角，在所引用出处《翻译学导论——理论与实践》一书中的译文名称是"信息类""表情类""操作类"，但为了与上文中语篇类型划分的称谓保持一致，此处改成了"信息型""表达型""操作型"。

三种类型,如图 1-1 所示。信息型文本主要包含新闻报道、商务信函、科技著作、政府公文、各种非虚构著作、商品使用说明、操作指令等,针对信息型文本的翻译,最核心的要求是将原文信息与内容原封不动地在译文中传达,语言表达上应多顺应目标语的表达习惯;表达型文本主要是指诗歌、戏剧、小说、散文等文学文本,这类文本中所蕴含的诗学范式、文体特征、艺术结构等美学特征往往要比文本内容与信息更重要,因而这类文本的翻译应竭力在译文中保留原文的诗学与美学特征;操作型文本主要包括商业广告、传教文本、政治宣传等,虽然也在向听众与读者传递信息,但这类文本更重要的目的是期望读者或听众能够采取行动,操作型文本的翻译重要的不是在译文中再现原文信息或保留原文语言特征,而是能够在译语读者中产生与原文相同的劝说效果,译文是否对原文进行增删、改写并不作为评判译文质量好坏的标准。此外,莱斯还补充了视听类语篇,主要是指借助语言之外的声音、图画等媒介进行交流的文本,如电影、电视等。这其实就是所谓的"多模态文本",对于这类文本的翻译,莱斯(2004)同样认为应将再现原文效果视为第一要求。

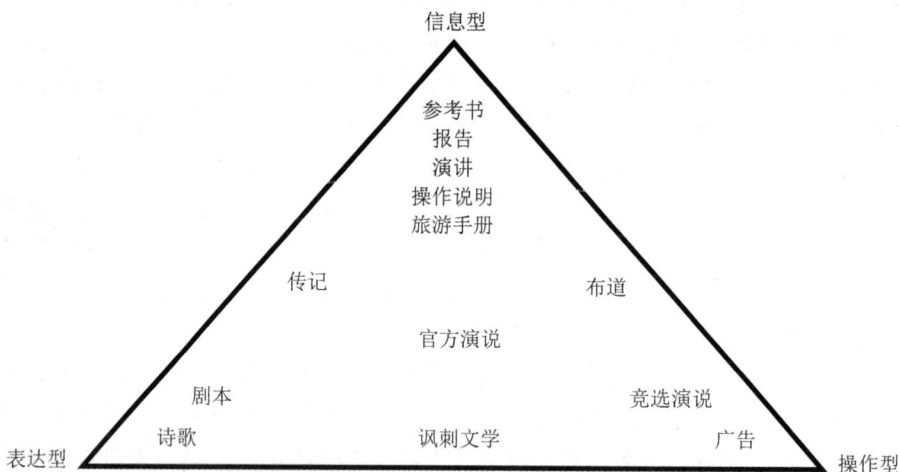

图 1-1　莱斯基于语言功能的语篇类型划分(芒迪,2007)

　　莱斯基于文本语言特征与文本功能的联系以及文本功能的不同侧重提出了更具针对性的翻译策略,在翻译界产生了很大的影响,也由此成为德国功能翻译学派的重要代表。当然,莱斯的论断也并非完美,不少学者针对其理论中的漏洞与不足进行了批评指正。其中,同为功能翻译学派的另一位著名学者克里斯蒂安·诺德(Christiane Nord)(2001)就指出,将语言的功能仅仅划分为三种是不全面的,有必要补充语言的第四种功能——寒暄功能,比如"你好""早上好"等打招呼的问候语，它承载着建立和维系交际双方社会联系的功能。彼得·纽马克(Peter Newmark)也在《翻译教程》(*A Textbook of Translation*，2001)一书中将文本功能进一步细分为六种，即表达功能、信息功能、呼吁功能、美学功能、寒暄功能和元语言功能，并在此基础上提出语义翻译法适用于表达型文本，交际性翻译法则适用于信息型文本和呼吁型文本。

此外，不同文本类型和相应语言功能之间也并非严格的一一对应关系，在实际翻译中经常需要加以辨别、灵活运用。以信息型文本中的科技文本为例，可粗略地分为硬科技文本和软科技文本。硬科技文本主要是指针对某一专业领域的科学技术专著，供相应科研人员进行技术分享与交流，其内容专业性较强，语言平实严谨，较少运用修辞手段；软科技文本主要是指科普类文章，目标读者是普通大众，为了便于读者理解，文中常用比喻、拟人等修辞性表达。也就说，不同类型文本之间的界限并非泾渭分明，往往呈现出一个渐变的状态。故而，玛丽·斯内尔·霍恩比(Mary Snell Hornby)在《翻译研究：综合法》(*Translation Studies: An Integrated Approach*，2001)一书中指出了不同文本类型之间的渐变状态，并给出了相应的翻译方法。

莱斯基于文本功能进行文本类型划分，提出了对应的翻译策略的思路，虽然其理论本身有诸多不足，也受到了学者们的诟病，但在翻译实践中依然带来了巨大的启示，让我们注意到文学文本、科技文本、广告文本等不同文本之间语言特征、语言功能的差异，并且意识到在翻译过程中译者要基于文本自身特点采用更具针对性的翻译策略。对于文学文本而言，审美功能居于首位，文体特征在译文中的再现比传递原文信息更重要；但对于科技性文本的翻译而言，译者更关注文本信息与文本内容在译文中的再现，原文文本的语言特征、表达方式和原文所表达的内容相比，居于次要地位。评判科技性文本译文质量的优劣也需要以译文是否忠实传达了原文本内容、是否实现了原文本意欲实现的信息功能为首要标准。

以上基于莱斯对语言功能、文本类型和翻译策略的探讨旨在从文本层面入手对翻译活动提供宏观的思考和方向指引，从而使译者在科技性文本翻译的过程中，一方面能基于科技性文本的语言特征和语言功能，以在译文中充分传递原文文本信息为第一要务，灵活运用直译、意译、编译、改译等翻译方法；另一方面，在科技文本翻译实践中，若面对意义与形式、内容与表达的两难选择，应牢记科技文本为信息型文本，并以传递原文文本信息与内容为首要责任，因而在翻译过程中应重表达内容与文本意义，轻语言表达与语言形式。

练习
practice

1. 莱斯的文本类型划分和我们平常所理解的文体类型有何异同？

2. 莱斯基于文本功能的文本类型划分对于翻译实践有哪些启示与指导意义？

第二节　翻译目的论

翻译目的论是近年来最受国内关注的西方翻译理论之一。一提到目的论，很多人立刻就会想到翻译过程中对原文的删减、改写，更有甚者将翻译目的论和"胡乱翻译"画上等号，翻译目的论似乎变成了为劣质译文寻找合理性依据的工具。翻译目的论所提出的根据翻译目的对原文进行增删，也很容易给初学者一种翻译可以不忠实于原文的错觉。实际上，翻译目的论是强调在翻译过程中，译者要关注原文功能、翻译目的、翻译委托三方面的内容，并据此在翻译过程中对原文进行调整、改写。

目的论的德文是 Skopos Theorie，原文来自希腊语 Skopos，意为"目的""目标""意图""功能"等，目的论的英文是 Skopos Theory。目的论最早由汉斯·弗米尔(Hans J. Vermeer)在《通用翻译理论基础》(*Groundwork for a General Translation Theory*)一书中提出，书中目的论的定义是："在目标语背景下，为译作目的和目标读者，译出译文。"(诺德，2001) 可见，翻译目的论强调关注译文的目的，译文目的决定了译者采用何种翻译策略和方法，从而提供可实现特定翻译功能的译文。因此，对翻译目的论而言，译者在翻译之前必须要搞清楚为什么翻译，译文文本的功能是什么。此外，为了更好地实现翻译目的，也需要了解译文读者的情况。

翻译目的论主张译本的预期目的决定翻译的策略和方法，并提出在翻译过程中必须遵循三大原则：目的原则(skopos rule)、连贯原则(coherence rule)和忠实原则(fidelity rule)。其中，目的原则是总原则，即翻译过程中所有翻译策略与方法的选择都是由翻译行为的目的而定。而翻译目的又分为三种：译者目的、译文交际目的和具体翻译方法所要达到的目的。连贯原则是指译文必须达到文内连贯，即译文必须符合目标语同类文本的行文规范。忠实原则是指译文与原文之间应符合互文连贯，即译文所表达的内容必须和原文内容相一致。值得注意的是，忠实原则居于目的原则和连贯原则之下。换言之，对于翻译目的论而言，相比译文对原文的忠实度，翻译目的实现居于首要地位，为了在翻译过程中实现翻译目的，会在一定程度上舍弃译文对原文的忠实度。这也正是翻译目的论最容易遭受误解之处。

误解背后的原因，很大程度上源自翻译目的论与国内翻译传统之间的差异。长期以来，国内对翻译的理解和对翻译原则的设定都是以原文为中心的。翻译过程中译文如何忠实再现原文是译者的不懈追求，以原文为中心引发了对翻译结果(译文)的重视。然而，翻译目的论却是以翻译目的和翻译委托为中心，将翻译视为"由目的所驱动、以结果为中心的交际活动"(芒迪，2007)。相较于对翻译的传统认知，翻译目的论更关注翻译的过程，认为翻译"不是简单地翻译词、句或是文本，而是引导潜在的合作，跨越文化障碍，促进功能性的交际"(芒迪，2007)。因此，翻译目的论更注重译文在目标语文化中是否实现了预期功能，而非译文与原文之间逐词、逐句对应的忠实关系。

本节用较大篇幅介绍弗米尔的翻译目的论，主要原因是其理论对翻译科技文本具有较强的指导意义。弗米尔的翻译目的论主要针对应用性文本，尤其是商业翻译，而当下科技文本翻译成为商业翻译的主要内容。在商业翻译活动中，译者必须遵守翻译委托，因为是否遵守翻译委托将直接决定译者的翻译活动是否得到认可并获得报酬，乃至译者的职业生涯是否能够持续。对于以科技文本为代表的应用性文本的商业翻译活动，既要实现原文文本的信息功能(忠实原则)，又要遵守译语文本规范(连贯原则)，当二者难以两全，译者必须作出取舍时，汉斯·弗米尔明确指出目的原则应当被置于首位。

汉斯·弗米尔的翻译目的论在产生重大影响的同时，也引发了不小的争议。诺德作为目的论第二代的代表人物，在对弗米尔等人的贡献给予肯定的同时也指出了其理论"只对非文学文本有效""对原文语言特色关注不够"等问题(芒迪，2007)。也就是说，诺德认为汉斯·弗米尔的目的论过于强调译文功能的实现，翻译过程中容易忽视原文语言特征在译文中的再现，因而她在弗米尔的理论基础之上做了进一步的完善。诺德在《翻译中的文本分析》(*Text Analysis in Translation*，1988)一书中勾画了一个更详细的功能模式，不但强调翻译过程中翻译功能的实现，还强调对文本结构、语言特征的分析与再现。她将文学文本的翻译也囊括在内，提出了两种基本翻译方法：文献型翻译(documentary translation)和工具型翻译(instrumental translation)。文献型翻译"充当原文化中作者与原文接受者之间交流的文献"，其中最典型的当属文学翻译。文献型翻译主要是指逐词翻译、直译和"译语情调翻译"，翻译过程中原文文化中的特有词汇往往会在译文中被保留下来；工具型翻译"充当目标语文化里新的交际行为中独立的信息传达工具，力图实现其交际意图，而又不让接受者意识到他们所读到或听到的文本曾以不同的形式在一个不同的交际情景中使用过"(诺德，2001)。换言之，译文读者在阅读由工具型翻译所产生的译文时，就像是阅读用他们自己的语言所写成的文本。

诺德(2001)在《翻译活动作为具有目的性的活动》(*Translating as a Purposeful Activity*)一书中提出了更灵活的分析模式，其中尤其突出三方面内容：翻译委托书的重要性、原文文本分析和翻译功能的排序。翻译委托书又称翻译指令，往往包含预期的文本功能、受众、文本接收的时间与地点、媒介、文本写作和翻译动机；原文文本分析包含文本主题、文本内容、翻译情景、文本结构、文本插图和排版、词汇特征、句子结构和语音特征；翻译功能排序包含判读译文的预期功能，为适应受众而需改变的功能成分、翻译风格和文本语言层面的处理。

诺德的翻译目的论虽试图将文学文本也纳入研究范围，但实际上她在《翻译活动作为具有目的性的活动》一书中所提出的相关主张仍然更适用于应用文本尤其是科技文本的翻译(诺德，2001)。全面深入了解诺德的翻译目的论，对于学习和从事科技文本翻译的译者颇具启发。它让翻译活动超越了仅仅关注文本层面忠实的局限，将翻译视为一种发生在具体社会语境下的社会活动，并深受文本之外的经济、读者接受度等因素的影响。翻译目的论提醒译者在翻译过程中不但要注重翻译文本和文本功能，还要关注译文受众与译语文化，并将翻译活动的外部语境等诸多因素都考虑在内。具体而言，主要体现在以下方面：

(1) 对于从事科技文本翻译的专业译员，不但要关注原文文本的信息功能及其在译文中的实现，更需要关注翻译委托人给出的翻译指令；

(2) 翻译过程中不但要考虑文本层面译文功能的实现，还应关注译文潜在读者的实际接受能力；

(3) 翻译过程中，还应从词汇、句法、篇章结构等层面对原文语言特征进行全面客观的分析，比较译文和原文在词汇、句法、篇章结构等层面的异同，并在翻译过程中作出相应调整。

练习
practice

1. 弗米尔提出的翻译目的论，主要有哪些特征？
2. 诺德的翻译目的论与汉斯·弗米尔的翻译目的论之间，有哪些异同？

第三节　科技翻译的译者素养和普遍方法

功能翻译理论与翻译目的论更多的是从宏观角度对翻译策略的探讨，强调翻译过程中文本类型的不同造成文本功能的差异，以及翻译委托人的翻译指令和译文目标受众对文本翻译策略可能造成的影响。然而在翻译实践中，经常会出现科技文本翻译的委托人并未在翻译合约中对译者提出特别要求，也并未要求译者基于译文功能而在翻译过程中大幅删减原文，更未要求采用改译、编译等方法，这时就需要译者采用传统的逐词逐句的翻译方法进行翻译。而且在编者看来，编译、改译也是建立在准确翻译基础之上的，只有对原文有全面准确的理解，才能在翻译过程中更好地删减冗余、保留必需，交出符合委托人要求的优质科技翻译译文。

因此，逐词逐句翻译仍然是初学者提升翻译水平的关键，学会逐词逐句翻译的一般方法，仍然是初学者应掌握的重要内容。在科技文本翻译过程中，初学者从一开始就应该关注科技译者素养、科技翻译标准、科技翻译过程等众多内容，并且从词法、句法和篇章三个维度关注科技文本的翻译方法与注意事项，力争从译者能力、译文评判和翻译过程三个方面对科技翻译有全面了解。此外，鉴于计算机技术在科技翻译中的广泛应用，如何使用机器翻译来提升翻译水平，也成为提高科技翻译水平的重要内容。总之，提升科技文本翻译能力需要掌握科技翻译的一般过程、常用方法和普遍原则。

科技文本翻译不但要求译者具有较强的语言能力，还需要对所译的科技领域有较深入的了解，这样才能保证理解原文无偏差，在译文中不说外行话。因此，科技文本翻译的译者应对所应遵循的职业规范、译者应具备的语言能力与科技素养必须具有全面的了解。此外，科技文本翻译在实践中具有任务重、时间紧的特点，借助计算机人工智能翻译软件可以极大地提高翻译的速度。因此，科技文本翻译过程中如何借助平行语料库和机器翻译做好译后编辑，都是当下从事科技文本翻译的译者所必备的技能。

科技文本翻译和文学文本翻译是不同的。为读者提供审美体验是文学文本所承载的重要功能；向读者传递知识与信息则是科技文本的首要功能。因此，科技文本翻译具有不同于文学文本翻译的标准与要求。从事科技文本翻译应遵守"忠实准确、通顺流畅、规范专业、简洁明晰"的翻译标准，这十六字方针不但指出科技翻译应忠实原文，译文流畅，还强调科技文本译文应遵循"专业""简明"这两条不同于文学文本翻译的标准。

学习科技文本翻译，还应关注翻译过程，并多向资深译者请教、学习。在较长一段时间里，国内很多翻译教程是不涉及翻译过程的，大多只关注翻译结果。译者对翻译活动的关注，也往往只从翻译结果入手，依据翻译标准对翻译结果作出优劣评价，忽视了对译者翻译过程中思维轨迹的探究，而恰恰是翻译过程中译者的思维活动对初学者有更大的启发与借鉴作用。好在近年来，翻译过程越来越得到翻译研究领域学者们的关注，眼动仪在口译教学研究中的广泛应用就是一个很好的例子。笔译过程和口译过程相比略有不同。实际上，较早的翻译过程研究也始于对笔译过程的探究。尤金·奈达(Eugene A. Nida)就曾在《翻译科学探索》(*Toward a Science of Translating*，2004)中将翻译过程明确划分为"理解—转换—表达"三个阶段，罗杰·贝尔(Roger T. Bell)在《翻译与翻译过程：理论与实践》(*Translation and Translating: Theory and Practice*，2001)中对奈达的模式进一步深化，将原文的理解和译文的表达又从句法分析、语义分析和语用分析三个维度进行了细化。了解专业译者在科技文本翻译中的思维过程，对初学者提升翻译技能大有裨益。

科技文本的翻译，具体而言主要涉及词法、句法和篇章三个维度。对这三个维度的深入分析对提高科技文本翻译水平具有重要意义。科技文本翻译过程中，无论是原文的理解还是原文信息在译文中的再现，都需要译者对科技文本有全面分析和深入理解，紧抓科技文本的语言特征，深入理解科技文本的内在涵义，而这一切最终都需要落实到文本的词法、句法、篇章层面，因而从词法、句法和篇章三个层面对科技文本翻译进行探讨对初学者具有重要意义。

从事科技文本翻译，还要有严谨的态度，要对原文和译文反复对照阅读，不放过任何一处细节。很多初学者在翻译结束后，很少愿意反复阅读译文，结果造成译文中总是含有蹩脚的表达。究其根本，在于译文中出现了很多"中介语"(interlanguage)，也就是介于原文和译语之间的语言表达。克服这一现象的最好方法就是反复诵读译文，再基于对英汉两种语言差异的掌握不断修改。初学者若能注意英汉两种语言之间的差异，并能对译文反复修改，则不啻为提升翻译能力的好方法。

最后需要指出的是，不少对科技翻译缺乏了解的人总是轻言"科技翻译易于文学翻译"，实际上，科技文本翻译的难度不亚于任何其他文类。而且，将西方先进的科学技术翻译到中国，有助于推动我国科技的不断发展与进步；把中国的优秀科技成果介绍给全世界，对提高我国科技发展的世界影响力意义重大。因此，青年学子要不畏艰难，努力学好科技翻译，促进中西方之间的技术交流，为我国的科技发展作出自己的贡献。

练习
practice

1. 相比文学翻译，科技翻译在翻译标准上有哪些不同？
2. 总体而言，科技翻译译者需要具备哪些素养？

第二章　科技翻译伦理

"翻译伦理"这一术语自 20 世纪 80 年代提出至今已有四十余年，其定义和内涵丰富多样。进入 21 世纪以来，随着语言服务业的发展，译者队伍迅速壮大，译者所应遵循的道德规范和行为准则成为学界研究的热点。进入新时代，精心构建对外话语体系、对外讲好中国故事的重要性更为凸显，翻译伦理被赋予了新的元素。

根据《新华大字典》和《现代汉语词典》，"伦理"是指人与人相处的道德准则；《朗文高阶英汉双解词典》(以下简称《朗文》) 将"伦理"(ethics) 解释为判断是非的道德准则；《牛津高阶英汉双解词典》(以下简称《牛津》) 将其解释为控制或影响人的行为的道德准则。《朗文》和《牛津》均将 professional ethics 作为短语列出，翻译为"职业道德"。由此可见，"职业伦理"和"职业道德"内涵相同。

中外学者对"翻译伦理"概念的界定较为多样。西方学者采用 ethics of translation、ethics of translating、translation ethics、translator ethics 等术语，其中较为明确的定义由安德鲁·切斯特曼(Andrew Chesterman)(2018)所提出：

"Translation ethics" (or "translator ethics") refers to the set of accepted principles according to which translation should be done (and, mutatis mutandis, interpreting), and hence the norms governing what translations should be like.

他认为翻译伦理(或译者伦理)是指一套关于应该如何翻译(或口译)的公认原则，以及由此形成的关于译作应该如何的规范。我国学界常采用"翻译伦理""译者道德"等术语。任文(2020)将翻译伦理定义为翻译行为主体在翻译(相关)活动中所应遵循的道德准则。杨荣广(2022)将其定义为译者为了实现某种善的价值取向所应遵循的伦理规范，具体表现为译者在翻译中所选择遵循的翻译行为规范和译作最后呈现的译文规范，其功能在于协调翻译活动利益相关者和相关要素之间的关系。

翻译伦理并非翻译(口译)活动的外在因素，翻译本身就是伦理行为。也就是说，译者的任何一个翻译决策都是在某个伦理原则或明或隐地牵引下所产生的行为。翻译伦理和翻译行为紧密相关，并不是抽离于翻译行为的抽象原则。

第一节　中国翻译伦理观

传统译论家最常谈到的两个字是"善"与"信"。三国时期，支谦(约3世纪)就在《法句经序》中提到"其传经者，当令易晓，勿失厥义，是则为善"。道安(314—385)坚持"委本从圣，乃佛之至诚也"，反对削胡适秦，饰文灭质，求巧而失旨。近代洋务运动时期的马建忠(1844—1900)在《拟设翻译书院议》中明确提出"善译"(中国翻译工作者协会，《翻译通讯》编辑部编，1984)：

夫译之为事难矣!译之将奈何？其平日冥心钩考，必先将所译者与所以译者两国之文字深嗜笃好，字栉句比，以考彼此文字孳生之源、同异之故。所有相当之实义，委曲推究，务审其音声之高下，析其字句之繁简，尽其文体之变态，及其义理精深奥折之所由然。夫如是，则一书到手，经营反复，确知其意旨之所在，而又摹写其神情，仿佛其语气，然后心悟神解，振笔而书，译成之文，适如其所译而止，而曾无毫发出入于其间。夫而后能使阅者所得之益与观原文无异，是则为善译也已……

这里的"善译"指的是翻译既要忠实于原作，又要让译文读者的感受等同于原文读者阅读时的感受。严复(1854—1921)在《天演论》译例言中提出"信""达""雅"。"善""信"二字，包含了中国文化中一种深刻的伦理精神。作为译者，应该与原文本身、原文作者、译文读者等为"善"，真正起到传达原文旨意，保证读者毫无障碍地理解原文的桥梁作用。在翻译的过程中，对原文作者的立意、原文的主旨要讲"信"，不歪曲原文作者的旨意，又要不负译文的读者，这是"翻译之道"，更是译者的"伦理之道"(彭萍，2013)。

我国翻译大家的伦理观还折射出高度使命感和爱国情怀。马建忠先生在提出设立翻译书院时，其中一个考虑因素为当时"学习西方"之急需：

窃谓今日之中国，其见欺于外人也甚矣。道光季年以来，彼与我所立约款税则，则以向欺东方诸国者，转而欺我。于是其公使傲睨于京师以陵我政府，其领事强梁于口岸以抗我官长。其大小商贾盘踞于租界以剥我工商，其诸色教士散布于腹地以惑我子民。夫彼之所以悍然不顾，敢于为此者，欺我不知其情伪，不知其虚实也。然而其情伪虚实，非不予我以可知也。外洋各国，其政令之张弛，国势之强弱，民情之顺逆，与其上下一心，相维相系，有以成风俗而御外侮者，率皆以本国语言文字，不惮繁琐而笔之于书。彼国人人得而知之，并无一毫隐匿于其间。中国士大夫，其泥古守旧者无论已。而一二在位有志之士，又苦于语言不达，文字不通，不能遍览其书，遂不能遍知其风尚。欲其不受欺也得乎？……

今也倭氛不靖而外御无策，盖无人不追悔于海禁初开之后，士大夫中能有一二人深知外洋之情实，而早为之变计者，当不至有今日也。余也蒿目时艰，窃谓中国急宜创设翻译书院，……

《拟设翻译书院议》全文 2500 字左右，但"中国"出现 80 次，足见马建忠先生对国家安危的关注和对翻译救国的殷切之情。徐光启(1562—1633)先生是最早把欧洲近代自然科学介绍到我国的学者，他提出："欲求超胜，必须会通；会通之前，必先翻译。"严复先生是我国近代译坛最早且最具影响力的以译介西学来启蒙大众、救亡图存的译者之一，他以《天演论》发出"自强保种"的有力宣言。林纾(1852—1924)先生在《译林·序》中也表达了译书救国的主张"吾谓欲开民智，必立学堂；学堂功缓，不如立会演说；演说又不易举，终之唯有译书。"梁启超(1873—1929)先生也提出"处今日之天下，必以译书为强国第一义""国家欲自强，以多译西书为本"。鲁迅、瞿秋白等知识分子也都具有深深的爱国情怀并在其译著中多有表达。

当代翻译大家也十分关注译者的责任和义务。林语堂提出，译者的第一责任就是对原著(者)以及本国读者负责；郭沫若提出要唤醒译书家的责任心；傅雷不但提出"神似论"的翻译主张，更对译者提出"要以艺术修养为根本：无敏感之心灵，无热烈之同情，无适当之鉴赏能力，无相当之社会经验，无充分之常识(杂学)，势难彻底理解原作，即或理解，亦未必能深切领悟"的要求；巴金提出为读者负责的观点；杨宪益出于传播中国文化的使命感，翻译《红楼梦》以直译为主、意译为辅。

进入 21 世纪，翻译伦理学研究日渐系统化，从宏观(人类命运共同体)、中观(行业和社群)和微观(译者)三个层面研究翻译中的伦理问题，强调译者的社会责任和价值观，主张以"真诚"代替"忠实"，实现"真"与"善"的统一。在这一观点下，"真诚"既是翻译活动的开端，也是整个翻译过程得以实现的保证。"真诚"意味着尊重他者，愿意倾听，愿意与他者交流，也愿意承担责任，并能以开放的胸怀对待源文本、他者、文化差异以及自身理解及诠释的局限性。这种坦诚的态度有助于建立译者的个人信誉，整个行业的信誉，甚至有助于减少文化冲突，助力人类命运共同体的构建(陈志杰，2021)。

练习 practice

1. 简答题。

(1) 康德说："世界上有两样东西，越反思越觉得它们令人赞叹与敬畏，这就是：我头上的星空和我心中的道德法则。"请阐述译者所应遵循的道德法则。

(2) 请查阅并阐述鲁迅先生有关翻译伦理的论述。

2. 用现代文解释彦琮法师所著《辩正论》中的"八备"，说明其核心思想，指出哪几条与译者的道德品质有关。

诚心爱法，志愿益人，不惮久时，其备一也。将践觉场，先牢戒足，不染讥恶，其备二也。筌晓三藏，义贯两乘，不苦暗滞，其备三也。旁涉坟史，工缀典词，不过鲁拙，其备四也。襟抱平恕，器量虚融，不好专执，其备五也。耽于道术，淡于名利，不欲高衒，其备六也。要识梵言，乃闲正译，不坠彼学，其备七也。薄阅苍雅，粗谙篆隶，不昧此文，其备八也。

第二节　国外翻译伦理观

法国当代翻译理论家安托瓦纳·贝尔曼(Antoine Berman)(1984)最早提出"翻译伦理"的概念。他认为翻译行为的"正当伦理目标"就是尊重原作，尊重原作中的语言和文化差异。他的翻译伦理目标实际上是通过传介"他者"来丰富自身。

美国解构主义翻译理论家劳伦斯·韦努蒂(Lawrence Venuti)(2002)提出在翻译中保存原文中语言文化差异的"异化"策略，其翻译伦理目标是促进文化更新和变化，反对文化殖民主义，反对"同一化"。韦努蒂倡导的是一种"存异伦理"思想，实际上是为弱势语言和文化请命。

安东尼·皮姆(Anthony Pym)(1997)提出译者的"文化间性"(interculturality)概念。他认为翻译是为了促成合作，译者不是在单一文化，而是在两个或两个以上文化交界处的文化交互空间(intercultural space)翻译，也就是说，译者具有文化间性。皮姆提出了以下五条翻译伦理：

(1) Translators are responsible for their product as soon as they accept to produce it.

译者一旦决定翻译，就应对其译文负责。

(2) Translators are responsible for the probable effects of their translations.

译者对其译文可能产生的结果负责。

(3) Translator ethics need not involve deciding between two cultures.

译者伦理不包括在两个文化间决策优劣。

(4) Transaction costs should not exceed the total benefits ensuing from the corresponding cooperative interaction.

翻译成本不应超过合作带来的收益。

(5) Translators, insofar as they are more than simple messengers, are responsible for the capacity of their work to contribute to long-term stable, cross-cultural cooperation.

译者不仅传递信息，还应致力于促进长期稳定的跨文化合作。

在众多主张理论中，影响最大的当属芬兰学者安德鲁·切斯特曼(Andrew Chesterman)提出的五大翻译伦理模式。切斯特曼总结出自20世纪80年代以来出现的四种翻译伦理模式，即再现伦理(an ethics of representation)、服务伦理(an ethics of service)、交际伦理(an ethics of communication)和基于规范的伦理(norm-based ethics)。但这四种模式互有矛盾，应用范围也不尽相同，基于此，切斯特曼提出了第五种模式，即承诺伦理(an ethics of commitment)(方梦之等，2014)。

再现伦理是指译者要忠实地再现原文或原文作者的意图，突显忠实(fidelity)与真实(truth)，体现对作者的尊重。服务伦理是指译者要符合客户要求，即符合与委托人协商后达成的要求。服务伦理是将翻译视为一种商业服务而提出的，主要是指翻译的功能模式，即翻译最终要符合客户要求，实现由委托人与译者共同商定的目标，关键是要在规定的时限内提供让客户满意的产品。服务伦理追求忠诚(loyalty)价值，这里所说的译者的忠诚首先是针对委托人的要求而言的，当然还包括对目标语读者及原作者的忠实。交际伦理认为译者应发挥跨语言交际的桥梁作用，体现对文化他者的尊重，满足不同社会间跨文化交往的需求。基于规范的伦理是指译者的翻译活动应符合目标语社会的各种规范，满足目标语文化的期待。承诺伦理认为译者应追求卓越，履行职业道德规范。

这些伦理规范仅为一般性原则，译者在翻译过程中应根据实际情况进行调整，使之具体化为特定的翻译策略和语言表达形式。此外，这些伦理规范适用范围不同，在具体的翻译活动中可能会发生冲突，译者需要根据具体的社会、历史情境进行权衡。

练习 practice

1. 简答题。

(1) 如何理解皮姆所提出的第三条翻译伦理"译者伦理不包括在两个文化间决策优劣"？

(2) 简述切斯特曼的五个翻译伦理模式对你的启示。

(3) 在翻译中，面对委托人的不正当要求，应如何处理？

2. 阅读下面的翻译小故事，并从翻译伦理角度讨论 Stefan Moster 的翻译行为。

A literary translator, Stefan Moster, translates a Finnish novel by Arto Paasilinna into German. At one point in this fantastical story set in Finland's pagan past, a semi-divine hero is born, who will save Finland from the threat of the new Christian religion. The hero is born on April 20. But the German version says he is born on April 19. Why? Because April 20 was Hitler's birthday. Many German readers will know this, so there is a risk that the novel will be taken as neo-Nazi propaganda, and this is a risk the translator refuses to take. On his own responsibility, he changes the date. He informs the German publisher, but not the author. Asking the author's permission would have run the risk of being refused, and this risk too he did not wish to take. Later, he makes the reasons for his decision public.

第三节　科技译者职业道德准则

2001 年国际翻译家联盟列出译者必须遵守的职业道德：保密、公正、按时、尊重客户、公平交易；谢绝自己所不能胜任的翻译任务；职业翻译工作者必须承诺不断更新自己的知识和技能。

美国翻译家联盟 2022 年审核通过的七条译者职业道德规范为：恪守专业精神，诚实守信，保护协会及行业声誉；翻译履历真实，翻译能力满足翻译任务要求；若有利益冲突或潜在利益冲突，告知相关方，并避免可预知的利益冲突；遵守契约；保守秘密；根据具体情境，准确、恰当、不偏不倚地传递意义；

持续提高知识、技能和能力。

中国翻译协会 2019 年发布的《译员职业道德准则与行为规范》规定了职业译员在从事翻译工作时应遵循的职业道德准则和行为规范，包括端正态度、胜任能力、忠实传译、保持中立、保守秘密、遵守契约、合作互助、妥用技术以及提升自我。

科技翻译因其科技属性、跨专业性等特点，对译者在以下方面要求尤为严格：

(1) 科技译者应秉持专业精神，恪守职业道德，诚实守信，自觉维护翻译行业声誉。比如，提高自身翻译能力和科技素养；慎重选择翻译策略；发现译文错误时，应及时、妥善地处理；发现源语信息不清楚或存疑时，例如存在歧义、事实性错误、术语不准确等问题时，应积极主动与委托方沟通，要求委托方进行解释、澄清或重新措辞以确保翻译质量；提高对源语信息的判断能力，拒绝非法或损害国家利益以及包含虚假、错误信息的翻译任务；热爱翻译事业和行业，对有损翻译行业声誉和尊严的不当言论和行为予以理性、正面的回应。

(2) 科技译者应具备科技翻译能力，且只从事自己有能力胜任的翻译工作。科技翻译专业性强，译者应对自身的科技知识和能力做出合理判断，确保翻译任务的难度在自己能力范围之内；应明确拒绝超过自身承受能力的翻译工作，确保能够按照约定条件保质保量地完成任务。

(3) 科技译者应忠实、准确地翻译源语信息，避免根据自己的意愿或观点进行修饰或增删等更改。科技翻译对准确性要求极高，一旦有错后果极为严重，小到让委托方蒙受损失，大到损害组织、国家的利益，因此翻译时应慎而又慎。但需要指出的是，翻译过程中也不能对语言文字进行生硬机械地转换，应以专业素养确保译文良好的可读性。

(4) 科技译者应严格遵守保密原则。译者应对委托方(用户个人)的隐私和机构信息、翻译内容严格保密。未经允许，译者不可自行通过任何渠道发布相关资料、照片、活动细节等信息，否则轻可导致损失，重可触犯法律。比如，若委托方为科技论文或著作作者、专利申请人，泄露信息可能损害委托方的知识产权；若委托方为商业机构，泄露信息可能导致委托方蒙受经济损失。此外，保密为译者责任，译者不应以保密为条件，向翻译活动参与方索要额外利益，不应利用翻译过程中所获得的保密信息获取个人利益。

(5) 科技译者应持续关注科技领域发展动态和翻译领域最新成果，养成终身学习的习惯，不断提升科技素养和翻译素养。科技发展日新月异，唯有持续学习才能培养科技敏感度，积累科技知识和常识，使翻译更准确、更专业。随着人工智能的发展，翻译技术功能越来越强大，比如德国 DeepL 公司开发的

DeepL 翻译器和人工智能公司 OpenAI 开发的 AI 聊天机器人 ChatGPT，都曾经或正在引起翻译行业的轰动。译者应妥善使用新技术来提升翻译效率，保证翻译质量。但需要指出的是，译者应对机器翻译译文进行译后编辑，不可将机器翻译的译文作为成品直接交付，除非和委托方另有协定。

特别需要注意的是，对翻译文本质量及客户负责是翻译职业道德的基本要求，但当基本层次的伦理目标与更高层次的伦理追求发生冲突时，应以"高"为上。比如，当客户要求或翻译内容会威胁更大群体的生命安全，导致他人的合法权益遭受巨大损失时——特别是这些负面效果还可能借由机器翻译等技术手段加速传播和扩大影响——译者在自身认知能力可合理预见的范围内对现有职业伦理规范的背离就可能具有道德的正当性。此外，不能因为翻译记忆和语料库等技术模糊了"译员"的概念就无视著作权/版权等问题，更不能因为翻译技术可以给译员带来诸多便利就将人类翻译的创意、人际交流的温度让渡给机器。即便将来翻译技术能够处理所有类型的文本，当待译文本的内容可能对未来社会和遥远世界带来负面影响时，译者应拒绝技术，作出不译的选择(任文，2019)。

练习
practice

阅读国际翻译家联盟欧洲区域中心发布的译者职业道德准则，找出它和中国翻译协会《译员职业道德准则与行为规范》的共核部分。

1. GENERAL CONDUCT

1.1 Basic principles

Translators and interpreters shall observe the respective laws and regulations. They shall always seek to maintain the highest standards in their work and serve their clients in the best possible way. In their relations with clients, with each other and with the public at large, they shall at all times act in a manner that does not bring the profession into disrepute.

1.2 Responsibility

Translators and interpreters have sole responsibility and liability for their work; any exclusion of liability has to be expressly agreed in writing. Translators should consider taking out professional indemnity insurance.

Translators and interpreters shall not knowingly mistranslate or misinterpret. Instructions received from the client cannot justify deviation from this Code.

1.3 Impartiality

Translators and interpreters shall carry out their work with complete impartiality and not express any personal opinions in the course of the work.

1.4 Confidentiality

Translators and interpreters shall maintain complete confidentiality at all times and treat any information received in the course of work as privileged, except when the law requires disclosure. They shall ensure that any person assisting them in their work is similarly bound. This confidentiality requirement continues beyond the respective assignment and also applies vis-à-vis persons who have acquired knowledge of the relevant information from another source.

1.5 Exploitation of knowledge acquired

Translators and interpreters shall not derive any gain from privileged information acquired in the course of work undertaken. In particular, they shall not interfere in a client's business relations with his/her customers.

2. QUALIFICATIONS

2.1 Competence

Translators and interpreters shall only work in languages and subject areas for which they are qualified and have the requisite skills. Translators should translate only into their mother tongue, their language of habitual use or a language in which they have proven equivalent competence.

2.2 Self-development

Translators and interpreters shall keep up to date with developments in the profession and the relevant tools by means of continuing professional development.

2.3 Titles

Translators and interpreters shall only use academic or other titles which they are permitted to use by law.

3. RELATIONS WITH FELLOW TRANSLATORS/INTERPRETERS

3.1 Competition

Translators and interpreters shall refrain from unfair competition (e.g. predatory pricing) and from public attacks on the reputation and competence of other translators and interpreters. Any criticism of another translator's or interpreter's work must first be expressed directly to the person concerned as objectively as possible.

3.2 Advertising

Translators and interpreters shall not infringe accepted advertising standards, nor make claims which cannot be substantiated. They may mention a client as a reference only with his/her prior consent.

3.3 Collegiality

Translators and interpreters, especially those belonging to professional associations, should assist each other where practicable. If offered work they are unable to accept, they should seek to recommend to the client another translator or interpreter who has the necessary skills.

3.4 Partners and employees

Relations with partners or employees are also governed by this Code where applicable. Inter alia, these persons must be granted reasonable remuneration for the services rendered.

4. PERFORMANCE OF WORK

4.1 Acceptance

Translators and interpreters are free to accept or turn down work, subject to any legal constraints. They shall decline work if it results in a clash of interests, if they believe it is intended for illegal or dishonest purposes or if they know that their own capacity, working conditions or time will prevent its proper completion. When declining work, they shall do so without unnecessary delay.

4.2 Contracts

Translators and interpreters should always seek to sign a written contract in advance of an assignment. The contract should, inter alia, specify the deliverables, deadlines, quality assurance methods, copyright, confidentiality, ownership of any translation memories used, terms of payment and access to background information.

The client must be informed as early as possible if it becomes evident that an agreed deadline cannot be met. If facing insoluble difficulties, the translator shall advise the client promptly so that they can jointly decide on how to proceed.

A fixed quotation in writing shall only be made if the full scope of the work to be performed is known. It shall explicitly state that services not specified in it will be charged at current rates.

第三章　科技译者素养

译者素养(translator competence)也称译者能力，这一概念由德国建构主义翻译学者吉拉里(Don Kiraly)在 1995 年基于心理语言学揭示的翻译心理过程中提出。与偏重语言能力的翻译素养(translation competence)概念不同，译者素养这一术语强调的是专业翻译者任务的复杂性以及所需的非语言技能，突出了翻译人员在现代翻译市场中作为专业人员所需要的专业能力。诺德(2018)在谈到译者的能力时，认为作为译者需具有获取专业知识能力、沟通能力、跨文化能力、写作能力、媒体运用能力、良好的通识教育研究能力、完美的抗压能力以及自我价值的主张能力等。译者素养的培养不能仅仅局限于源语言知识和目标语知识的培养，它还是一个包括知识(语言知识、文化知识、翻译技术、学科领域知识、伦理知识及其他与翻译相关的理论或事实知识)、能力(源语言与目标语语言能力、信息技术能力、翻译管理能力等)和特质(细节关注力、自我反省意识、乐于合作等)的综合体。

科技发展和经济全球化使中国对外交流与合作日益频繁。随着综合国力日渐增强，我国在全球科研领域扮演的角色也越来越重要。科技发展离不开科技信息的传播与交流，坚持"引进来"和"走出去"相结合，不仅要把国外先进的科学技术和科技成果介绍到国内，还要把我国的优秀科技成果推向全世界，推动中国科学技术的国际传播，成为科技强国。作为科技交流的桥梁，科技译者必须适应国家战略和经济社会的发展，积极服务国家对外开放战略，结合专业方向进行国际交流、传播中国声音，这是翻译人员肩负的历史使命。

要履行这一使命，一名合格的科技译者除了扎实的语言功底、广博的科技知识，还需要具备丰富的翻译理论素养和实践素养，以及良好的信息素养。

第一节　语言素养

译者的语言素养体现在源语应用能力、目标语应用能力和双语转换能力三个方面。

翻译是对两种语言深层次的理解和再造，翻译过程就是对原文进行剖析之后使用译语文字将其表达出来的过程。良好的双语言功底是从事翻译的基础。茅盾先生认为"精通本国语文和被翻译的语文是从事翻译的起码条件"。译者水平的高低，首先取决于他(她)对于原作的阅读理解能力。同时，译者熟练驾驭目标语，才能自如而确切地表达。

　　想要驾驭英文不仅要学会英文单词、短语，懂得句型、语法，更重要的是能够熟练掌握并灵活运用英文的表达方法，能够用英文思维，用英文写作。单其昌(1999)在《汉英翻译技巧》的序言中提出："如果一个人英文水平低，别的条件都具备，他只不过具备了全部翻译条件的百分之三十，还是不能进行翻译的。反之，如果一个人的英文水平较高，那么他就具备了翻译条件的百分之七十，也就是说，英文水平是搞好汉英翻译的关键。要搞好汉译英，就必须在英语方面打好基础。"

　　除了扎实的英文基础，译者还必须具有较好的母语修养。现在，社会上不同程度地存在重外语、轻中文的倾向。想要成为合格的译者必须在努力提高外文水平的同时，下功夫提高自身的中文素养。母语水平欠佳可能导致无法准确理解甚至根本看不懂原文，翻译也就无从做起了。在汉译英翻译实践中，译者若对中文原文理解不到位，严重依赖汉英词典或机器译文，在取词和表达上便会遇到困难，无法将中文的原意正确地表达出来。比如：

　　这一发现颠覆了人们对全球农业状况的认知。

　　This discovery subverts people's cognition of the state of global agriculture.

　　此译文将"颠覆"和"认知"分别取词，生硬照搬汉英词典上这两个中文词的对应英文，分别译为 subvert 和 cognition，割裂了"颠覆对……状况的认知"的搭配含义，明显不符合原义。根据上下文，"颠覆"应为"彻底改变"，而不是字面含义"采取阴谋手段从内部推翻"；"认知"应为"(片面或错误的)认识"之意，并非是心理学术语。这种情况下，可行的策略是查询英文搭配词典或语料库，使用较明确的宾语"认知"的备选译文 assumption、awareness、understanding 等，查询与其搭配的常用动词，或查询英文同义词近义词词典，挑选动词 change 的近义词中具备"彻底改变(想法)"含义的词汇，从中选取最符合本句意思的备选项。经过查询，可以对译文作出如下改进：

　　This study punctures assumptions about the state of global agriculture.

　　改进的译文中 punctures assumptions 准确表达了本句中"颠覆"与"认知"的内涵，在语体和语域方面也和原文保持了一致。

　　这个例子同时也说明，在翻译实践中译文的质量与译者的双语转换能力密切相关。良好的汉英转换能力要求译者不仅要精通汉语和英语的词汇和语法，还要熟悉中国和英语国家文化背景知识以及英汉两种语言在语音、词汇、句法、修辞以及使用习惯上的种种差异，以便将规范通顺的汉语转换为规范通顺的英语，而不是生硬牵强的"中式英语"。同时，在转换过程中，科技译者也应具有文化意识。所谓文化意识，就是译者要认识到翻译既是跨语言，也是跨文化的信息交流，而文化差异与语言差异一样可能成为交流的障碍。缺乏文化意识的译者可能只顾及字面上的转换，而忽视了语言背后的文化问题导致误译。只有对双语文化都有坚实的知识储备，才能在翻译"文化词语""文化现象"时，最忠实地传达原文的意思，呈现出准确精彩的译文。

那么怎样才能夯实基础进而增强语言基本功和双语转换能力呢？首要的方法就是大量阅读，只有平时大量地"输入"，需要"输出"时才能信手拈来。其次，多读原版外文书，增加词汇量，积累地道的英语表达方法，提高语言敏感性和欣赏能力，积累知识的同时也巩固了语言学习。再次，还可以通过报纸、杂志、电视、互联网等多样化渠道了解西方国家的政治、文化、经济等方面的知识来开阔视野，参加一些活动来增长见闻。最后，可进行规范系统的写作训练，尤其是英文写作训练(包括文学写作和实用写作)，不仅要掌握各类文体的写作规范、流程和技巧，还要特别注意汉英两种语言在这些方面的差异。

第二节　科技素养

语言知识是做好科技翻译的基础，而专业技术知识则是必要条件。科学研究在宏观领域和微观领域同时深入发展，一方面，研究视角日益广阔，综合、跨学科前沿学科层出不穷，另一方面，学科越分越细，研究也不断深入。科技译者的工作对象日益精深、细分、综合，因此，译者的科学素养和专业知识储备的作用更加凸显。如果专业知识储备不足，就会遇到翻译质量的瓶颈，译文的关键信息就不够准确和专业，甚至出现误译。因此，对于英文专业背景的科技译者，像过去那样边实践边积累专业知识已远远不够。科技发展的形势要求译者在掌握通用翻译能力的基础上必须进行系统的"二次学习"，尽可能储备充分的专业知识以应对工作中的挑战。

科技翻译对专业性、科学性和准确性要求很高。缺乏相应的科技背景知识，翻译过程中很容易断章取义或望文生义，比如把"锂电池寿命"译为 lithium battery life。如果译者对锂电池原理有所了解的话就会知道，真正的锂电池是用金属锂或锂合金作为负极，危险性大，很少应用于日常电子产品。目前所说的锂电池通常是指"锂离子电池",对应的术语是 lithium-ion battery 或 li-ion battery,它是把锂离子嵌入碳(石油焦炭和石墨)中形成负极,较为安全。人们通常称其为"锂电池",因此容易引起混淆。另外,锂离子电池的寿命通常指的是其循环寿命,对应的术语是 cycle life,其单位通常为充电周期数(如 500 次),而非小时或年。techtarget.com 网站给出的 battery life 定义是 a measure of battery performance and longevity, which can be quantified in several ways: as run time on a full charge, as estimated by a manufacturer in milliampere hours, or as the number of charge cycles until the end of useful life。由此可见,在英语中,battery life 有多重含义,对应的汉语可以是"电池续航""电池电量"和"电池循环寿命"。因此,为了避免歧义,汉语语境中的"锂电池寿命"最好译为 lithium-ion battery cycle life 或 li-ion battery lifetime。

科学技术发展日新月异,新的专业知识不断涌现。译者应树立终身学习的意识,持续跟踪国际、国内科技发展,广泛了解和学习各科技领域的专业词汇和基本知识。在此基础上结合个人禀赋,重点深入掌握某些特定领域作为自己日后进行科技翻译的主攻方向。在科技翻译过程中要秉持严谨认真、科学准确的态度,具备锲而不舍的毅力,养成查阅文献和多次检查确认的习惯,精益求精,确保翻译质量。

第三节 翻译理论与实践素养

从宏观意义上说,翻译理论是关于翻译活动的知识。它描述和分析翻译活动中的各种现象,并针对有关现象提供一系列的观点和诠释,包括一般原理与具体的指引、建议和启示。从这个角度来看,翻译理论属于基础性的研究,重在解释翻译的性质、功能、双语转换机制、语言与思维的关系等问题,同时还借鉴语言学、文艺学、符号学、社会学、哲学、美学、心理学、文化学、交际学、信息论等学科的理论来解释种种翻译现象。从微观意义上说,翻译理论涉及如何根据不同的语篇类型、翻译目的以及译文读者对象而采取不同的翻译策略和方法。可见,微观意义上的翻译理论比宏观翻译理论的涵义更狭窄和具体一些,它的作用也更为具体。例如,确认和解释翻译中的问题,指出解决问题所涉及的所有因素,建议适用的翻译策略和标准等。

从理论与实践的辩证关系而言,翻译理论的职能和作用主要表现为三个方面(刘宓庆,1999)。首先,翻译理论具有认知职能(cognitive function)和启蒙作用。翻译理论是对翻译活动现象和规律的认识、描述、分析和探讨。通过翻译理论的揭示，我们可以认识到翻译这一特殊语言活动的实质、翻译活动的各项基本规范以及翻译过程中的行为模式。其次，翻译理论具有执行职能(performing function)和指导作用。翻译理论能够指导我们认识翻译活动的规律，并凭借有关的理论论证和方法论的引导在实践中有选择地"实施"理论所提供的多种参照性对策，进而使翻译技能从自在行为上升为自为行为，使翻译实践成为高层次的语际交流，而不仅仅是匠人式的技艺。再者，翻译理论具有校正职能(revising function)和规范作用。这一职能源于翻译理论的执行职能。翻译理论提供了多种参照性对策，使我们在翻译实践过程中能够有选择地采用有效的实施手段，同时也使我们更能辨明正误，校正偏差。

翻译实践与翻译理论之间的关系是一个需要正确对待的问题。一个翻译能力强的人一定具有较强的语言能力，而一个语言能力强的人不一定具有较强的翻译能力。认为只要具备了足够的语言能力就能做好翻译，从而得出翻译理论对实践的指导意义不大的结论，这其实是对翻译的一种极大误解。克里斯蒂娜•谢芙娜(Christina Schäffner)(2000)提出，在学习翻译的开始就应该学习什么是翻译以及翻译能力应该包括哪些成分等翻译专业知识。诚然，在中外翻译历史上的确有一些翻译家没有接受过任何翻译培训，也不懂翻译理论，但却能做好翻译。其实并不是他们没有翻译理论指导，而是他们在翻译中会自觉或者无意识地运用到某一翻译理论。如果他们能够清晰地认识指导自己翻译实践的理念及其所属的理论范畴，或许会使自己的翻译实践更加理性，知道自己为什么会采用某种翻译策略，其含义是什么，后果又是什么，让自己的译作更加完美(穆雷，2012)。

翻译理论并不能把劣译者打造成翻译高手，也不能让学翻译的人变得聪灵或有鉴赏力。翻译理论的价值在于总结归纳各类有效的翻译策略和技巧，发现翻译活动过程的规律，解释翻译活动中出现的各种现象，并为翻译实践提供不同的操作途径。古人云："授人以鱼，不如授人以渔。"懂一些翻译理论，能使译者掌握翻译规律并在其指导下进行翻译实践，提高翻译能力，对翻译问题知其然也知其所以然，达到事半功倍的效果。

因此，译者应学习以下基本的翻译理论知识：一是有关翻译活动本身即翻译本体的知识，主要包括翻译本质、翻译过程、翻译单位、翻译标准以及翻译的思维方法等；二是关于翻译主体即译者的知识，主要包括译者在翻译中的主体地位、译者在翻译过程中的创造性和局限性、译者的知识结构、译者的能力构成以及译者的职责和道德素质、译者的常见错误分析等；三是关于翻译客体也就是文本的意义或信息的知识，主要包括意义的定义、意义的种类、文化差异、思维差异以及语言差异对翻译意义的影响、意义的可译性与不可译性、各种不同意义的特点以及与翻译的关系等。马会娟(2013)根据应用翻译的成果，提出在以发展翻译能力为目标的翻译教学中，学习者应该掌握翻译的性质、翻译的类型、翻译的标准和原则、翻译的目的、翻译的方法、翻译的步骤、文本类型与翻译方法、职业翻译的特点或因素、译员的职业素质、翻译项目管理、翻译的伦理等十一项内容。

翻译理论研究历来多以文学翻译为蓝本，但研究取得的很多成果完全可以也应当为科技翻译提供借鉴和指导。从实践层面上讲，科技译者可以将一些文学翻译的技法引入，使译作逻辑严谨又充满语言的美感，向读者更畅通地传达科学信息。从理论层面上讲，文学翻译理论的一些核心观点也同样适用于科技翻译，例如采用优化的译语表达方式、发挥目标语优势、实现准确和通顺相统一、寻求意义对等而非形式对应等观点。如果不去主动了解这些精辟的论述，译者在实践中可能就要花费更多精力去自行探索或多走弯路。优秀的译者应该具备理论素养，广泛汲取杰出翻译理论家的智慧，继而逐渐形成自有的翻译策略，有效指导科技文本的翻译实践。通过充分实践，译者可以形成独到的理论心得，丰富科技翻译学的知识宝库。

第四节 信息素养

信息素养指"能够认识到何时需要信息，能够检索、评价和有效利用信息，并且对所获得的信息进行加工、整理、提炼、创新，从而获得新知识的综合能力"(陈坚林，2010)。

传统的文学翻译更强调译者的语言文学功底、阅读理解能力及表达能力，而且通常任务量不大，形式比较单一，时效要求也不是很强。而今，信息化给翻译工作带来了巨变。我们正在经历从文本翻译到超文本翻译、从线性翻译到非线性翻译、从静态翻译内容到动态翻译内容的快速转变过程。越来越多的软件、网站、游戏、手机 APP 等需要推向全球市场，语言服务和本地化的需求空前增长，多样化和专业化的趋势日益凸显，待译内容和格式越来越复杂。翻译工作不仅数量巨大、形式各异，且突发任务多、时效性强。与此同时，计算机辅助翻译技术、语料库技术和机器翻译技术飞速发展，"译前编辑—机器翻译—译后编辑"已成为很多译者的工作模式，大幅提升了工作效率。同时，本地化工程和技术工具，如 Alchemy Catalyst、SDL Passolo、Adobe RoboHelp、WebWorks Publisher 等已经在本地化服务客户方和服务商中得到了普遍应用。利用新技术分担大量重复性工作，使译者得以专注于文本的重点和难点，从而实现更好更快的翻译要求。这些变化都对译者的信息素养提出了更高的要求。社会变迁和技术进步是无法改变的时代潮流，译者唯有努力调整自身才能应对新挑战。科技译者必须全面提高自身的信息素养，借助各种信息化技术，提升翻译工作的效率，保证译文的质量。

科技译者的信息素养主要体现在以下几个方面。

1. 计算机辅助翻译技能

计算机技术和辅助翻译工具的应用能力已经成为现代翻译职业人才的必备素质。据《中国地区译员生存状况调查报告》(中国科学院科技翻译协会、传神联合信息技术有限公司，2007)的统计，61%的译员在使用辅助翻译工具，80%的译员在使用在线辅助参考工具，可见翻译职业化进程对译员的计算机辅助翻译(CAT)工具应用能力要求之高。当前各大语言服务公司对翻译人员的招聘要求中都强调会熟练使用 CAT 或本地化工具。在现代化的翻译项目中，翻译之前需要进行复杂文本的格式转换(如扫描文件转 Word)、可译资源提取(如提取 XML 中的文本)、术语提取、语料处理(如利用宏清除语料噪声)等；在翻译过程中需要了解 CAT 工具中标记(Tag)的意义，掌握常见的网页代码，甚至要学会运用 Perl、Python 等语言批处理文档等；翻译之后通常需要对文档进行编译、排版和测试等。可见，计算机相关知识与 CAT 技能的高低将直接影响翻译任务的进度和翻译质量。

2. 信息检索能力

在信息化时代，知识正在以几何级数增长，新的翻译领域和专业术语层出不穷，再聪明的大脑也难以存储海量的专业知识。如何在有限的时间内从浩如烟海的互联网中找到急需的信息，如何通过专业语料库验证译文是否准确地道等，都要求译者必须具备良好的信息检索、辨析、整合和重构的能力。

翻译任何专业的内容都需要对该专业有一定的了解。在笔译当中，大约七成的时间都是在查阅文献、学习背景知识、确认自己的译文是否符合该领域的表达习惯。由此可见信息检索能力的重要性。在开始翻译前，检索搜集相关文献可以帮助译者尽快地掌握相关领域的知识和术语表达，为准确地翻译打下基础。在翻译过程中，译者如果在表达方面遇到问题，可以利用网络或者电子词典查找目标语的相关文本，学习表达方式和语言风格，从而提高译文的质量。

当代科技译者尤其应该熟练掌握主流搜索引擎(如百度、必应、Semantic Scholar 等)以及专业语料库的特点与用法、查询关键词的选择、检索语法的使用等，以提升检索速度和检索结果的质量。同时，译者还应熟悉互联网上丰富便利的翻译相关资源，如专业翻译论坛(Proz、Translationdirectory、Translatorscafe 等)、知识问答平台(知乎、百度知道、Wordreference、Answers 等)、各种百科全书资源(百度百科、维基百科、大英百科全书、Encyclopedia 等)。

3. 术语能力

术语能力是指译者能够从事术语工作、利用术语学理论与术语工具解决翻译工作中术语问题所需的知识与技能，具有复合性、实践性强的特点，贯穿于整个翻译流程中，是翻译工作者不可或缺的一项职业能力(王少爽，2011；2013)。术语管理是译者术语能力的核心内容，已成为语言服务中必不可少的环节。目前流行的 CAT 平台大都提供了术语管理系统(TMS)，便于用户管理和维护翻译数据库，提升协作翻译的质量和翻译速度，促进术语信息和知识的共享，传承翻译项目资产等。因此，当代译员需具备系统化收集、描述、处理、记录、存储、呈现与查询等术语管理的能力(冷冰冰、王华树、梁爱林，2013)。

4. 译后编辑能力

机器翻译在信息化时代的语言服务行业中具有强大的应用潜力，目前几乎所有主流的 CAT 工具都可加载机器翻译引擎，如 Google Translate、DeepL Translator、有道翻译、腾讯翻译君等。智能化的机译系统可帮助译者从繁重的文字转换过程中解放出来，工作模式转为译后编辑(post-editing)。2010 年，翻译自动化用户协会(TAUS)对全球语言服务供应商的专题调研表明，49.3%的供应商经常提供译后编辑服务，24.1% 的供应商拥有经过特殊培训的译后编辑人员，其他则分发给自由译者。目前译后编辑已成为译员必备的职业能力之一，当代译员需要掌握译后编辑的基本规则、策略、方法、流程、工具等。

练习
practice

1. 译者的语言素养主要体现在哪几个方面？
2. 译者如何提高自身的科技素养？
3. 科技译者应该掌握哪些翻译理论知识？
4. 科技译者的信息素养主要包含哪几个方面？

第四章　科技翻译标准

　　翻译是一项有目的的活动，要根据社会、文化、交际等因素选择适当的翻译策略或方法，向客户交付符合特定需要的产品。翻译可以分为两个不同的领域：文学翻译和非文学翻译。文学翻译涉及想象中的个人、自然、人类居住的星球，非文学翻译涉及知识、事实和思想、信息、现实；文学翻译强调的是价值和风格，非文学翻译强调的是事实和信息的清晰性。

　　国内外各种翻译理论和翻译流派提出的翻译标准数不胜数。清末翻译家严复提出了著名的"信、达、雅"标准，即译文的思想内容忠于原文，译文表达通顺，译文风格优美。林语堂提出了"忠实、通顺、美"的翻译标准；傅雷主张"形似神似"；钱钟书提出了"化境"的翻译标准；王佐良认为"一切照原作，雅俗如之，深浅如之，口气如之，文体如之"。英格兰翻译家亚历山大·弗雷泽·泰特勒(Alexander Fraser Tytler)(2007)提出"翻译三原则"：译文应该完全摹写原文的思想；译文的风格和写作方式应该与原文相同；译文应该与原文的行文一样自如。美国翻译家尤金·奈达(1964；1969)则提出"动态对等"或"功能对等"的翻译标准，他指出翻译不应拘泥于原文的语法结构和字面意义，而应该着重于译文读者的反应，使译文读者获得与原文读者相似的心理感受，这就要求译文在词汇意义、文体特色等各个层面上尽可能与原文保持一致，成为原文的"最近似的自然对等语"。

　　科技文体旨在阐述客观事物的本质特征，描述其发生、发展及变化的过程，集中体现语言的信息功能，其文本呈现的客观意义比其他文体更强，较少涉及个人情感和社会文化背景因素。科技翻译隶属于非文学翻译，因此上述的翻译标准对科技汉英翻译实践具有一定的指导意义，但针对性还不够。基于科技文体具有的表述客观、逻辑严密、行文规范、用词正式、句式严谨等特点，译者在翻译时应更多关注信息是否准确传递，术语表达是否地道准确，译文是否通顺流畅、简洁明晰等。基于此，我们将科技翻译的标准定位为忠实准确、通顺流畅、规范专业、简洁明晰。"忠实准确"体现的是原文和译文信息的一致性，"通顺流畅""规范专业""简洁明晰"则体现出译文的可读性和可接受性。

第一节 忠实准确

科技翻译的任何错误，哪怕仅仅是不准确，都可能给科学研究、学术交流、生产发展等带来不良影响甚至巨大损失，因此，科技翻译的第一条标准就是"忠实准确"。译文必须忠实于原文，必须客观、准确、完整地表达、传递、重现原文的内容。原文和译文的指称意义必须一样、完全或者基本重合，和原文表达的是同一概念，描述的是同一过程或者变化等。译者必须充分理解原文所表述的内容，包括对原文的词汇、语法、逻辑关系和科学内容作出精准理解和忠实传递。

【经典赏析】

原文：多数网络运营商都有专用网络，通常针对企业客户，所有专用网络都由交换中心的特殊设备提供，但是用户所在地和交换中心之间的传输要么使用铜线接入网络，要么借助光纤电缆的容量(如果已有光纤电缆或专门提供电缆)。

译文：Most of the network operators have specialized networks, usually aimed at the business customers. All of these networks comprise special equipment located in exchange buildings. However, the transmission between the customer's premises and the exchange building uses either the copper access network or the capacity in optical fiber cables—if these cables already exist or a cable is provided specially.

赏析：原文中有很多补充信息，例如"通常针对企业客户"被译为 usually aimed at the business customers，并用逗号隔开，作为非限定性定语，与原文保持高度一致。原文括号中的"如果已有光纤电缆或专门提供电缆"作为"光纤电缆"的补充信息，在译文中采用了放在句尾加破折号补充说明的方式。整体译文句式结构运用得当，信息准确，高度忠实于原文。

【案例解析】

◇ 例1

原文：当然，实际操作中，电话系统可能还需要包含一个机制，让说话者向听话者示意其说话意愿，因此，我们需要在图 1.1(b)的组件中加入一个铃铛，并让其在电路中与电源和开关相连。

原译：In practice a telephone system needs to include a mechanism for the caller to indicate to the recipient that he wishes to speak. Therefore, we need to add a bell associated in a circuit with a power source and a switch to Fig. 1.1 (b).

解析：原译中丢失了中文原文的"当然"一词，对 needs 和 wishes 等词的时态使用没有顾及原文中略带虚拟的含义，"组件"一词也漏译，同时 add 后面的整体结构略失平衡。译文中增加了 of course 和 assembly，并把虚拟含义也加入其中，更忠实于原文，表意更加精准，add 后面也采用了分隔修饰使结构更加平衡。

译文：In practice, of course, a telephone system would need to include a mechanism for the caller to indicate to the recipient that he wished to speak. Therefore, we need to add to the assembly in Fig. 1.1 (b) a bell associated in a circuit with a power source and a switch.

◇ 例2

原文：在实际电话网络中，最重要的是将连接每个用户的成本降到最低，因为一个电话网络中包含几千个甚至几百万个用户，所以只要提供给每个用户的设备数量稍有减少，整体成本就会大幅降低。

原译：In a practical telephone network, the most important is to minimize each customer's cost. Since there are many thousands or millions of customers on a telephone network, reduction in equipment for each customer would result in large overall cost savings.

解析：原译将"最重要的"译为 the most important，结构不完整，译文根据上下文补上了 requirement 一词，更忠实于原文的实际含义。原译将"连接成本"直接省译为 cost，丢失了部分含义。原译未能译出"稍有"之意，译文在 reduction 前加入 any，精准表达原意。原译将"提供给每个用户的设备数量"译为 equipment for each customer，丢失了"数量"和"提供"的含义。

译文：In a practical telephone network, the most important requirement/what is the most important is to minimize the amount of cost associated with connecting each customer. Since there are many thousands or millions of customers on a telephone network, any reduction in the amount of equipment needed to be provided for each customer would result in large overall cost savings.

◇ 例3

原文：尽管图 1.4 这种架构对仅有几部电话的网络来说十分合适(许多小型公司

和家用电话系统确实使用的是这类设计），但是它不好扩大规模。

原译：Whilst the arrangement shown in Fig. 1.4 is quite practical for networks of just a few phones and many small office and household telephone systems are based on such designs, it is hard to scale up.

解析：原文中将"许多小型公司和家用电话系统确实使用的是这类设计"放入括号，说明其意思并不完全与前面的句子并列，而仅仅作为额外补充，原译将其作为并列句用 and 连接不完全忠实于原文。译文采用将其放于破折号后的方式处理，与原文使用括号有异曲同工之妙，既保证了信息准确，又忠实于原文。

译文：Whilst the arrangement shown in Fig. 1.4 is quite practical for networks of just a few phones—indeed, many small office and household telephone systems are based on such designs—it is hard to scale up.

◇ 例 4

原文：如果手机在通话时移动到小区边界处，基站的信号就会减弱，这时如果控制器同时检测到该信号和来自相邻小区基站的更强的无线电信号，两个基站之间就会进行"交接"，交接时不会干扰通话。

原译：If the handset during a call travels towards the boundary of its cell, and if the weakening signal from the base station and the strengthening radio signal from the adjacent cell's base station are detected by the controller, the two will exchange the signal connection without interruption to the call.

解析：原文中"在通话时"应该是"移动"的状语，因此原译将其放在 handset 之后作定语不妥。原文中"交接"加了引号，暗含拟人之意，翻译时采用有类似词义的 handover 一词更加契合，从内容和文体上都更忠实于原文。

译文：If the handset travels towards the boundary of its cell during a call, and if the weakening signal from the base station and the strengthening radio signal from the adjacent cell's base station are detected by the base station controller, a "handover" between the two will be managed without interruption to the call.

◇ 例 5

原文：数据与语音在其性质和来源上都有本质不同，主要区别在于：数据是一种以字母和数字形式呈现的信息，这些字母和数字由计算机等设备生成、存储并展示；而语音是由人类产生的声音。

原译：Data is different from voice in its nature and origin. The primary difference is that data is information represented in letters and numbers which are generated, stored and displayed by computers; whereas, voice is generated as a sound by humans.

解析：原译中未能译出"本质不同"中的"本质"一词，"以字母和数字形式"漏译了 the form of，"计算机等设备"则仅译出了"计算机"，未完整表示"等设备"。相比之下，译文把所有的细节信息都准确译出，更加忠实于原文。

译文：Data is fundamentally different from voice in its nature and origin. The primary difference is that data is information represented in the form of letters and numbers which are generated, stored and displayed by devices, e.g. a computer; whereas, voice is generated as a sound by humans.

第二节　通　顺　流　畅

"通顺流畅"指的是译文文本符合目标语的语法、句法和表达习惯，文理通顺，结构合理，逻辑关系清楚，语言流畅自如，容易为读者所理解和接受。具体来说，译文的遣词造句和语篇构建简洁明了、重点突出、语义流畅、通顺易懂，翻译时应当避免因为语言形式的差异而影响信息的理解，避免死译、硬译、语言晦涩难懂的现象，避免出现翻译腔等问题。

【经典赏析】

原文：从此以后，电话技术就成为人类交流的普遍手段，人们利用亚历山大•格雷厄姆•贝尔提出的原理在全世界建立了电话网络。

译文：Since then, telephony has become the ubiquitous means of communicating for humankind, and telephone networks using the principles of Alexander Graham Bell have been implemented throughout the world.

赏析：本句在翻译时，基于逻辑关系，将"利用亚历山大•格雷厄姆•贝尔提出的原理"转为 telephone networks 的后置定语 using the principles of Alexander Graham Bell。第二句也没有选择"人们"来作主语，并列句中的主语从 telephony 到 telephone networks，避免了主题跳跃，使译文过渡自然，更加通顺流畅。

【案例解析】

◇ 例1

原文：大多数现代汽车的设计都是相对符合空气动力学的，这使得它们能够以最小的阻力穿过空气。

原译：The designs of most modern cars are relatively aerodynamic, and this makes the cars able to pass through the air with minimal resistance.

解析：原译按照中文字面顺序翻译，主语为 design，使得"它们"译为 them 指代不清，译为 the cars 又不够简洁流畅。"这使得"翻译为并列句略显冗长。译文将主语改为 Most modern cars，第二个分句译为分词短语或定语从句，可使译文更加简洁流畅。

译文：Most modern cars are designed to be relatively aerodynamic, allowing/which allows them to pass through the air with minimal resistance.

◇ 例2

原文：通过木门、水体的振动可以传播声波，当有人敲击栏杆远端或者远处火车到来时铁轨上产生的振动也可以传播声波。

原译：Sound waves can be carried as vibrations through a wooden door or water, and when someone at the far end is banging a railing or as the distant train approaches the railway lines, vibrations can also carry sound waves.

解析：原文中给出了传播声波的四种振动场景，原译将其分为前后两句话，且因后两种情况需要用句子表示，故采用了 when 引导的状语从句，最后则采用了不同于第一句被动结构的主动结构，句式整体跳跃，不够通顺流畅。译文将其合为一句话，使用 through…as well as through…将几种情况并列起来，并将后两种复杂情况用带有从句的名词结构表示，整体表达更加简洁流畅。

译文：Sound waves can be transmitted through the vibrations of wooden doors and water bodies, as well as through the vibrations of the rails when someone knocks on the far end of the railing or when a distant train arrives.

◇ 例3

原文：波形(可能为长度、电压、功率等)振幅中连续的波峰或波谷之间的时间称为"周期"，周期包括波形的一次完整循环。

原译：The waveform might be length, voltage, power, etc. The time between successive peaks or troughs in the waveform amplitude is known as the "period", and a period comprises one full cycle of the waveform.

解析：原译将"波形"及括号中的内容单独译为一句话，最后一句也单独翻译，用 and 与前句相连，虽然意思完整，但表达略显烦琐，表意不流畅。译文将 waveform 调整为整句主语的最后一个词，将括号中的内容仍置于括号中，用定语从句表示，放在 waveform 之后，这样表达更加顺畅。最后一句转为定语从句，补充说明前面的 period，表意更加连贯。

译文：The time between successive peaks or troughs in the amplitude of the waveform (which might be length, voltage, power, etc.) is known as the "period", which comprises one full cycle of the waveform.

◇ 例 4

原文：有线电视网络能够通过网络传输的一个备用电视频道向 ISP 提供高速数据链路，这一技术需要通过所谓的"电缆调制解调器"实现，它将计算机的输出转换为可被电视转播系统兼容的信号。

原译：Cable TV networks can go through one of the spare TV channels carried through the network to provide a high speed data link to an ISP, the technique requires a so-called cable modem, and it converts the output of the computer into a signal compatible with the TV distribution system.

解析：原译中按照中文句式将英文译为三个并列句，第一句将"通过"这一表示方式状语的短语译为主动词，"提供"这一动作转为目的状语，表意重心发生了转移，且后面两句话结构松散，翻译不够通顺。译文用 over 构成的介词短语表达"通过"之意，表意重心仍为"有线电视网络能够提供高速数据链路"，后两句单独成句，并利用后两句首尾相接的特点将第二句转为定语从句，整体翻译更加流畅。

译文：Cable TV networks are able to provide a high speed data link to an ISP over one of the spare TV channels carried through the network. The technique requires a so-called cable modem, which converts the output of the computer into a signal compatible with the TV distribution system.

◇ 例 5

原文：第十一章介绍电信网络的其他模型，并对网络结构这一概念进行详细分析。

原译：Chapter 11 introduces other models of the telecommunications networks, and the concept of network architecture is examined in more detail.

解析：原译将原文译为并列句，两个分句采用了不同的主语，意思跳跃，不甚连贯顺畅。译文以第一句为主要信息，采用被动句的形式，将 Chapter 11 置于句尾，第二句采用非限定性定语从句的结构跟在后面进行补充说明，表意中心突出，意思连贯，通顺流畅。

译文：Other models of the telecommunications networks are introduced in Chapter 11, where the concept of network architecture is examined in more detail.

第三节　规 范 专 业

　　"规范专业"指的是译文所涉及的科学技术领域的专业术语表述符合科技文体和术语的规范。科技文体是应用于科学技术领域进行沟通和交流的正式文体，其读者通常具有相当的认知能力和教育文化水平，并熟知其所在行业的相关专业术语和惯用语。科技文体的正式程度越高，专业术语、定义和概念也就越多，专业化和行业化程度也就越高，对译文的规范化、专业化要求也就越高。在科技翻译中要达到"规范专业"的要求，译者应当具备一定的科学技术知识，或者通过查询专业资料和求教于相关领域的专业人士，透彻理解原文内容，尽可能译出规范、专业、约定俗成的定义、概念和术语等。

【经典赏析】

原文：通常而言，这些服务会将用户拨打的号码转换为另一个号码以完成通话，还会执行特殊的收费方案(如由受话方而非主叫方支付)以及费率——"0800"号码就是一个例子。

译文：Typically, these services are based on the translation of the number dialed by the subscriber to another number in order to complete the call, together with some special charging arrangements (e.g. the recipient rather than the caller pays) and tariffs—as is the case of "0800" services.

赏析：原文中的"用户""转换""收费方案""受话方""主叫方""费率"等属于通信服务领域的专业词汇，分别采用了 subscriber、translation、charging arrangements、recipient、caller、tariffs 等规范表述，而非 user、change、charging plan、the person answering the phone、the person calling、fee 等意义相近但缺乏专业性的表述，整句也因为采用了这些专业词汇

的规范表述而显得更为正式。

【案例解析】

◇ 例 1

原文：话务员服务网络提供了通话援助、紧急情况服务、号码查询、残障人士
　　　特殊援助服务等。

原译：The operator-services network provides call help and access to the urgent
　　　services, telephone number consulting, disabled special help services, etc.

解析：因为行业知识欠缺，原译把"援助"译为常见的 help，"紧急"译为 urgent，
　　　"号码"译为 telephone number，而非 assistance、emergency、directory
　　　等专业词汇。这样的翻译显得随意，缺乏规范，很难为专业人士所接受。

译文：The operator-services network provides call assistance and access to the
　　　emergency services, directory enquiry, special assistance services for the
　　　disabled, etc.

◇ 例 2

原文：需要注意的是，这三种网络本质上使用的是同一种交换系统，但移动网
　　　络会使用无线电链路和移动性管理，而非 PSTN 和有线电视网络所使用
　　　的有线接入方式。

原译：It is noted that all three networks use essentially the same type of changing
　　　system, but the mobile networks use radio links and mobility management
　　　in place of the wire connection of the PSTN and Cable TV networks.

解析：原译把"交换系统""有线接入"等专业术语按照字面意思译为 changing
　　　system 和 wire connection，译文则采用更为规范的表述 switching system
　　　和 fixed-wire access，表述更加正式规范。此外，"需要注意的是"译为
　　　it is to be noted that 更为准确，按照语法规范，but 之后的并列主语从句
　　　的引导词 that 应保留。

译文：It is to be noted that all three networks use essentially the same type of
　　　switching system, but that the mobile networks use radio links and mobility
　　　management in place of the fixed-wire access of the PSTN and Cable TV
　　　networks.

◇ 例 3

原文：整合的行为在其他几类网络中也很常见，如货运轮船需要整合集装箱，

即在世界各地的重要港口将集装箱重新装载到其他货轮上。

原译：This organizing act is also common in many of the other types of networks, such as the organization of boxes on ships carrying products, which are put onto other ships at important ports around the world.

解析：原文中的"整合""货运轮船""集装箱""重新装载"等都属于海洋运输行业的专门词汇，不能简单译为 organizing、ships carrying products、boxes、put 等口语化的解释性词汇，而应该采用更为规范的译法 consolidation、cargo ships、containers、re-stack 等。"重要港口"使用 strategic 比 important 更符合港口特有的战略地位的描述，专业词汇的使用让整句话表述更加专业规范。

译文：This act of consolidation is also common in many of the other types of networks, such as the consolidation of containers on cargo ships, which are re-stacked onto other ships at strategic ports around the world.

◇ 例4

原文：就每个电路的单位成本而言(前提条件是如果使用了容量)，由于传输链路系统的容量越大，其成本效益越高，因此实际传输系统通常采用多路复用设备。

原译：Since the cost benefit of a transmission link system is higher the larger its capacity—in terms of the unit cost per circuit (the pre-condition is that the capacity is used)—practical transmission systems usually connect multiplexing equipment.

解析：原文中的"成本效益"属于经济学术语，所以不能按照字面简单翻译为 cost benefit，而应该采用专业译法 cost effectiveness，括号中的"前提条件是如果使用了容量"可以采用英语的"provided + 从句"来表示，显得更加简洁正式，原译中"采用"一词的译法不够规范。

译文：Since the cost effectiveness of a transmission link system is higher the larger its capacity—in terms of the unit cost per circuit (provided the capacity is used)— practical transmission systems usually adopt multiplexing equipment.

◇ 例5

原文：在实际系统中，在标尺上应用大量的量化电平就能将量化噪声控制在较低水平。

原译：In real systems, the quantization noise is kept pretty low by using a large number of quantization levels on the ruler.

解析：原文中的"实际"指的更多是"实践"而非"真实"之意，因此使用 practical 更为规范，"较低水平"指的是技术领域中普遍认可接受的水平，使用口语化表述 pretty low 有失专业，acceptably low 则更为规范。

译文：In practical systems, the quantization noise is kept acceptably low by using a large number of quantization levels on the ruler.

第四节 简 洁 明 晰

"简洁明晰"是科技文体的重要特点和基本要求之一，简言之，就是译文要简短精练、一目了然，尽量避免译文生硬啰嗦、冗长烦琐、赘词太多、不得要领、不必要重复等。

【经典赏析】

原文：虽然数字技术有其优点，但由于本地铜线回路具有高覆盖率、高成本效益和高适应性，大多数 PSTN 用户线路仍然使用模拟信号。

译文：Despite the merits of digital working, the ubiquity, cost effectiveness and adaptability of the local copper loop mean that the majority of PSTN subscriber lines are still analogue.

赏析：原文有三个句子，中间夹杂了让步、因果等逻辑关系，译文采用 despite 介词短语替代"although + 从句"表示的让步关系，将"具有"这一动词去掉，原来的三个"高"直接略去，只保留了"覆盖率""成本效益""适应性"等这些本地铜线回路的特点作为主语，添加 mean 一词，将原本的因果关系从句转为简单句，最后一句将"使用模拟信号"直接转为 are analogue，整体译文用词、句式简洁明晰，表意忠实准确。

【案例解析】

◇ 例 1

原文：各种解决方案都为静态存储的数据提供保护措施。某些解决方案只是保护磁盘，但 MQ Advanced 可保护消息数据本身。即使磁盘被黑客入侵，该磁盘上的所有内容也不易受到攻击。

原译：Various solutions offer protection for data at rest. Some solutions only protect the disk, but the MQ Advanced protects the message data itself.

Even if the disk is accessed by hackers, all the contents on that disk are not vulnerable.

解析：原译中，第二句重复了 solutions 一词，因第二句的"某些解决方案"属于第一句的"各种解决方案"的一部分，所以可采用省略的形式直接用 some 作为主语。"被黑客入侵""所有内容"均按照字面生硬翻译，不如使用 hack 和 everything 简洁。最后一句中 that disk 指的是同一张磁盘，因此可以采用代词 it 替代，这样句子整体用词简洁、表意明晰。

译文：Various solutions offer protection for data at rest. Some only protect the disk, but the MQ Advanced protects the message data itself. Even if the disk is hacked, everything on it is not vulnerable.

◇ 例 2

原文：工业物联网(IIoT)把机器、云计算、分析和人员整合到一起，这样就可以提高工业流程的性能和生产效率。工业公司可以借助 IIoT 实现流程数字化，转变业务模型，提高性能和生产效率，同时还能减少浪费。

原译：Industrial IoT (IIoT)brings machines, cloud computing, analytics, and people together, and therefore it can improve the performance and productivity of industrial processes. Industrial companies can use IIoT to realize the digitalization of processes, transform business models, improve performance and productivity, and at the same time decrease waste.

解析：原译中，第一句被译为两个并列分句，并用 therefore 表示因果关系，结构略显松散，可用不定式短语替换第二个分句。第二句中的"实现流程数字化"可以直接将"数字化"作为动词更加简洁，"同时还能减少浪费"在原译中为介词短语＋动词短语，与该句中其他动词结构并列，不如使用"while ＋ -ing"形式简洁。整句经过修改之后，逻辑关系清晰，用词更为简洁。

译文：Industrial IoT (IIoT) brings machines, cloud computing, analytics, and people together to improve the performance and productivity of industrial processes. With IIoT, industrial companies can digitize processes, transform business models, and improve performance and productivity while decreasing waste.

◇ 例 3

原文：数字孪生是实体对象的虚拟模型。它可以跨越物体的生命周期，并使用物体上的传感器发送的实时数据来模拟行为并监控操作。数字孪生可以复制许多现实世界中的物体，小到工厂中的单台设备，大到完整的装置如风力涡轮机，甚至可以复制整个城市。

原译：A digital twin is a virtual model of a physical object. It spans the object's lifecycle and uses real-time data sent from sensors on the object to simulate the behavior and monitor operations. Digital twins can replicate many items in the real world, from single pieces of equipment to full installations in a factory such as wind turbines, and they can even replicate entire cities.

解析：原译中的前两句完全按照中文结构，分译为两个单独的句子，使得两句话长度不均衡，措辞不够简洁。可将第一句转为介词短语，把两句合二为一，句式结构更加简洁。第三句的主语跟第一句相同，原译由单数突然转为复数，破坏了流畅性。可采用代词 it 替代，前后衔接更好，用词也更简洁。"甚至可以复制整个城市"在原译中单独用一个句子表示，与前文用 and 连接，不够简洁。可将其与"from…to…"搭配在一起，变成"from…to…, and even to…"，意思更为连贯。

译文：As a virtual model of a physical object, a digital twin spans the object's lifecycle and uses real-time data from sensors on the object to simulate the behavior and monitor operations. It can replicate many real-world items, from single pieces of equipment in a factory to full installations such as wind turbines, and even to the entire city.

◇ 例 4

原文：智能网络(IN)由遍布全国的若干网络中心组成，中心内包括控制系统及数据库，能够提供各类先进的交换服务。

原译：The Intelligent Network (IN) comprises several centers around the country, these centers contain control systems and data-bases, and they can provide a variety of advanced exchange services.

解析：原译按照中文结构将句子译为三个分句，结构比较松散。可采用 containing 短语作后置定语和 that 引导定语从句的结构，将句子合为一个主句，这样修改以后信息重点突出，表述简洁明晰。

译文：The Intelligent Network (IN) comprises several centers around the country containing control systems and data-bases that provide a variety of advanced exchange services.

◇ 例5

原文：如前所述，模拟电波形或电信号被定义为对输入波形的模拟，这里的输入波形指的就是说话人声音导致的气压变化。

原译：As was previously discussed, an analogue electrical waveform or electrical signal is defined as an analogue of an input waveform, and the input waveform here refers to the air pressure variation caused by a speaker.

解析：原译中将"如前所述"译为 as 引导的从句，其中的 was 可省略。原译采用 A is defined as B 的句式，虽无错误但不够简洁，可将"被定义为"转为介词短语 by definition，句式变为更加简洁的 A is B 结构。原译最后一句采用并列句译法，不够简洁，可利用最后一句与前一句最后的 input waveform 之间的关系，将其译为同位语短语，表达更加精练，语义也更为明晰。

译文：As previously discussed, an analogue electrical waveform or signal is, by definition, an analogue of an input waveform, i.e., the air pressure variation caused by a speaker.

练习
practice

1. 简答题。
(1) 文学翻译和非文学翻译各有哪些特点？
(2) 如何理解科技翻译的标准？
2. 根据科技翻译的标准，赏析下列翻译。
(1) 望远镜分为两种：折射望远镜和反射望远镜。折射望远镜最先被发明出来，也是目前使用更为广泛的一种，而反射望远镜则是迄今为止最大的望远镜，但两种望远镜的基本原理是相同的。仪器的大透镜(镜头)在焦点处形成被观测物体的真实图像，然后利用原理基本类似放大镜的接目镜对图像进行观察和放大。

Telescopes are of two kinds, refractor and reflector. The former was first invented, and is much more used, but the largest telescope ever made is the latter. In both the fundamental principle is the same. The large lens, or mirror, of the instrument forms at its focus a real image of the object looked at, and this image is then examined and magnified by the eye-piece, which in principle is only a magnifying-glass.

(2) 集成数字 PSTN 的重要优势之一在于，与笨重的机电和半电子模拟交换相比，数字电子设备占地面积大大减少，从而节省了运营成本。

One of the major advantages of an integrated digital PSTN is operational cost savings resulting from the significant reduction in floor space taken up by digital electronic equipment compared to the bulky electro-mechanical and semi-electronic analogue switches.

(3) 因为存在消色差这一缺点，使用普通透镜(即经过视觉矫正的透镜)进行天文摄影的效果并不令人满意。

This imperfection of achromatism makes it unsatisfactory to use an ordinary lens (visually corrected) for astronomical photography.

(4) 这种情况与大型仪器在展示微小细节方面的优势有很大关系，再怎么增加小型望远镜的放大率，也无法与使用相同放大率的大型望远镜看到的物体一样清晰，当然，前提是大型望远镜的物镜在工艺方面完全相同，且空气的光学条件良好。

This circumstance has much to do with the superiority of large instruments in showing minute details. No increase of magnifying power on a small telescope can exhibit things as sharply as the same power on the larger one, provided, of course, that the larger object-glass is equally perfect in workmanship, and that the air is in good optical condition.

(5) 机器学习算法无须编程，可直接从采样数据(或称为训练数据)中建模来进行预测或作出决策。这些算法已广泛应用于医药、电子邮件过滤、语音识别、农业、计算机视觉等无法或者难以通过开发传统算法来执行所需任务的领域。

Machine learning algorithms build a model based on sample data, known as training data, in order to make predictions or decisions without being explicitly programmed to do so. They are used in a wide variety of applications, such as medicine, email filtering, speech recognition, agriculture, and computer vision, where it is difficult or infeasible to develop conventional algorithms to perform the needed tasks.

3. 赏析下述科技短文翻译中体现的翻译标准。

(1) 中国人民大学高瓴人工智能学院助理教授周骁列举了人工智能(AI)在日常生活中的几种使用情况。例如：由 DeepMind 在 2016 年左右开发的人机博弈型 AI 的阿尔法狗用的就是搜索算法、深度神经网络和强化学习；图像识别型 AI 则用于交通管理中的车牌识别、嫌疑人追捕中的人脸识别等场景，在 2020 年左右取得突破，精度达到了 99%。

Zhou Xiao, an assistant professor at the Gaoling School of Artificial Intelligence, Renmin University of China, listed a few examples of Artificial Intelligence (AI) that people can use in daily life. There are the gaming AIs that can compete with humans playing games, such as AlphaGo developed by DeepMind around 2016, which features a search algorithm, deep neural networks and reinforcement learning. There are also image recognition AIs that can recognize car plate numbers for traffic management and human faces in the hunt for suspects, which made breakthroughs around 2020 with their precision rate reaching 99 percent.

(2) 这几种 AI 都深刻改变了我们的生活。比如人机博弈领域，AI 在过去的五年中发展迅猛，象棋或围棋中的人机对战已成为提高技能的方式；人脸识别 AI 技能成熟后，手机"刷脸"也成为人们开锁或支付的方式。现在，随着生成预训练转换器 GPT 的快速发展，人机交互的方式可能会再次发生改变。

These applications have already launched a revolution in daily lives. In gaming, AIs have made major progress over the past half a decade, and people now can play games such as chess or GO with smart AIs as opponents so as to sharpen their own skills. Image recognition AIs have also become mature, and people could easily open a lock or pay for a deal by holding their smartphone in front of their face. Now with the GPT (Generative Pretrained Transformer) making fast developments, the way people interact with computers might change again.

(3) 百姓网创始人、CEO 王建硕试图用简单的语言解释 GPT 识别人类语言的原理，"这一算法好比在空间中确定一个点的位置，知道三个参数(坐标值)就可以确定三维坐标系中的一个点了，但在人类语言中，当我们说到某个物体时，可能需要上千个参数才能确定这一物体究竟是什么。比如苹果的参数有能吃、水果、颜色红或绿、在树上生长、一般直径不超过 10 cm 等等。参数越多，我们就越能精准确定某物体。"

"That algorithm works in a way like finding a point in a space," said Wang Jianshuo, founder and CEO of Baixing AI. He tried to explain how it works in a plain language: "With three parameters (coordinates), one can locate a point in a 3-dimensional coordinate system. In the human language one might need thousands of parameters to describe an object; for example, an apple needs parameters such as edible, fruit, green or red in color, grown from a tree, and generally smaller than 10 centimeters in diameter to be defined. The more parameters that can be defined, the more accurately the AI can find the right point."

(4) 当苹果和香蕉同时出现时，能吃、甜、水果等参数就不够了，光凭这三个参数，不管是人还是 AI 都没有办法确定描述的是苹果还是香蕉。这时候就得增加参数，比如形状是长条还是类似圆形，颜色是红、绿还是黄，我们才能进行更精准的判断。"这就是 GPT 的基本算法，本质上是建了一个上千参数的模型来定义物体，其实人脑也是这样操作的，只是我们不会感受到罢了。"

With parameters such as "edible", "sweet" and "fruit", neither humans nor an AI could distinguish apples from bananas. But with the parameter concerning the shape included, namely "long" or "round", and with the parameter of the color being red, green or yellow, one could be surer about his/her judgment. "That's also how GPT works—most of the times it takes thousands of parameters for the AI to define an object like our human brain does, and we just do not realize that we are doing it," he said.

(5) 据复旦大学计算机科学技术学院教授张军平表示，ChatGPT 中语言模型的数据量尤为关键，ChatGPT 大概用了 285 000 个 CPU 和 10 000 多颗 A100 型 GPU、其训练文本数据达到 45 TB，才取得了此次的突破。ChatGPT 主要受益于能够使用语言模型基于大规模数据来训练庞大神经网络模型的大型语言模型。"ChatGPT 会生成与用户意图相匹配的多个回合的响应。ChatGPT 捕获先前的会话上下文来回答某些假设问题，这大大增强了会话交互模式下的用户体验。"

"ChatGPT was based on the training of 45 Terabytes of data to have made its breakthrough, for which purpose it used about 285,000 CPUs and over 10,000 model A100 GPUs," said Zhang Junping, a professor on computer technology at the School of Computer Science, Fudan University, who stressed that the amount of data the language model involves is of key importance to GPT: "ChatGPT benefits mainly from Large Language Models (LLMs) that train huge neural network models with large-scale data using Language Models (LMs)." "ChatGPT generates responses that match the user's intent with multiple turns. ChatGPT captures previous conversational contexts to answer certain hypothetical questions, which greatly enhances the user experiences in conversational interaction."

第五章　科技翻译过程

为提高译文质量，避免科技翻译中出现问题，译者必须完善翻译的各个过程。总体而言，翻译过程包括"解码、转换、编码"。在这种模式中，译者先阅读理解源语言，之后在大脑中加工转换，最后输出表达为目标语。根据以上论述，翻译过程主要分为两个步骤。

科技翻译的第一步是理解。它包括两个部分：一是原文语言的理解，二是原文内在联系的理解。译者要分析原文特点，利用语言知识正确理解原文的语义。在理解原文过程中，译者还可以通过查找术语、查阅平行文本等方式来进一步提高准确性。

科技文本高度简洁，若因此导致句子内在联系不明，则需要译者仔细推敲理解，以避免语言啰嗦。比如，科技文本中为了使表达简洁，通常会使用大量代词，如果直接翻译这些代词，可能无法清楚表达所指代的内容。因此，在翻译中译者应该仔细分析，正确理解指代内容，并在译文中清晰体现。

科技翻译的第二步是表达。译者要保证译文措辞准确、表达通顺、语言简洁、体现科技文本的语言特点，遵循翻译目的，从而译出符合目标语规范的文本。

第一节　理　解　原　文

翻译的过程首先是对原文理解的过程。任何译者在拿到一篇文章时首先要面临的问题就是理解原文，理解和处理原文作者所表达的各种信息。可以说，理解是翻译过程的第一步，是表达的前提，是翻译成功的先决条件。

翻译中的理解比一般的阅读理解要更透彻、更细致，不仅要弄清原文词语的表面意义，寻找译文中的对等词语以进行一字一句地机械式转换，还包括分析表层、挖掘深层、揣摩风格等几个方面。

就科技翻译而言，译者既要理解内容又要研究形式，既要琢磨原文词法、句法和修辞特点，又要考虑作者背景、专业知识以及某些词语的特殊含义。

原文理解中应注意以下几点：

(1) 正确选取词义：翻译过程中选取词义并非简单依赖于各种词典，而是需要译者在仔细阅读和充分理解原文的基础上，根据原文语境来斟酌确定。

(2) 英文表达常常使用词汇、句子、语法等来实现语篇内各成分联结，而中文表达主要依靠意义上的逻辑关系。因此，译者翻译时需要注重形合与意合的转化。

(3) 翻译是一种跨文化语言交流活动。翻译过程中，译者要充分考虑文化差异，运用翻译方法和技巧来使译文为读者所接受。

【经典赏析】

原文：2023 世界机器人大会将于 8 月 16 日至 22 日在北京举办，机器人产业前沿成果和最新展品将集中亮相。

译文：The World Robot Conference 2023 (WRC 2023) will be held in Beijing from August 16 to 22, at which cutting-edge achievements and the latest robot industry exhibits will be unveiled.

赏析：通过理解可知两句话之间的内在逻辑关系，译文以 at which 将逻辑关系明确表达出来。

【案例解析】

◇ 例 1

原文：设计任何载人任务中返回地球的飞行路线，都需要在保证机组人员能够安全返回地球表面的同时，尽量减少反应堆因失误重新进入地球大气层的可能性。

原译：Designing a flight path back to Earth on any manned mission requires ensuring that the crew can safely return to the Earth's surface while minimizing the possibility that the reactor will re-enter the Earth's atmosphere by mistake.

解析：原译未能深入理解原文，按照汉语语序采用直译，导致原译中主语部分过长句子焦点不够清晰，建议使用形式主语 it。原译将第三个小句译为 while 引导的状语与原文意思有偏差，应忠实于原文，将第二小句译为状语，同时遵循英文表达习惯将状语后置，使句子主干更明晰。

译文：It could be important to design the return to Earth flight path on any crewed missions to minimize the probability of an inadvertent reentry of the reactors into the Earth's atmosphere while enabling a safe return to the Earth's surface for the crew.

◇ 例 2

原文：可以通过设计反应堆系统、设计航天器上船员舱和监测飞行器内部和周围的局部辐射场，包括自然辐射场和反应堆产生的辐射场，以及通过监测每个乘员的个人接触情况来实现。

原译：By designing systems for the reactor, designing the crew quarters on the spacecraft and monitoring areas of radiation inside and outside the spacecraft, whether natural radiation or radiation from the reactor, as well as monitoring individual radiation exposure of each astronaut, to realize this.

解析：原文是汉语中常见的无主句，这种句子的作用在于描述动作、变化等情况，而不在于叙述"谁"或者"什么"进行这一动作或发生这个变化。因此，在翻译时常常需要补充主语，在科技翻译中通常使用 it 或者 this。原译采用直译，将 by 引导的超长介词短语置于句首，显得头重脚轻，且句子无动词。建议将主句提前，令句子结构一目了然。

译文：This can happen/This can be achieved/realized through the design of the reactor system and the crew quarters on the spacecraft and through the monitoring of the local radiation fields in and around the vehicle, including natural radiation fields and those introduced by the reactors, as well as through the monitoring of each crew member's individual exposures.

◇ 例 3

原文：地球、木星和土星的磁场由一个几乎与旋转轴对齐的偶极控制，天王星和海王星和上述不同，其特点是一个大的偶极子之间的倾斜旋转轴和强劲的四极八极贡献内部磁场。

原译：Unlike Earth, Jupiter and Saturn, whose magnetic fields are controlled by a dipole almost aligned with the axis of rotation, Uranus and Neptune are characterized by a large dipole with a slanted axis of rotation and a strong fourth and eighth contributing internal magnetic field.

解析：原文的内容为天文学，所以在理解原文时要考虑专业背景知识，原译中"四极八极"译为 fourth and eighth，虽与中文对应，但是与作者想表达的"四极八极贡献内部磁场"内容并不一致，在天文学中有专业术语，应译为 quadrupole and octupole。

译文：In contrast to the magnetic fields of Earth, Jupiter and Saturn, which are dominated by a dipole nearly aligned with the rotation axis, those of Uranus and Neptune are characterized by a large tilt between the dipole and spin axis and strong quadrupole and octupole contributions to the internal magnetic field.

◇ 例4

原文：随着核热推进反应堆和系统设计的确定，需要进行深入的事件序列分析，包括全面探讨在启动、全功率运行、关闭和休眠期可能发生的始发事件，这是概率风险评估技术的一部分。

原译：As the design of the nuclear thermal propulsion reactor and system is determined, an in-depth event sequence analysis is required, including a comprehensive discussion of initiating events that may occur during start-up, full power operation, shutdown, and dormantic phases, as part of probabilistic risk assessment techniques.

解析：原译按照汉语的语篇结构进行翻译，即先交代细节，然后逐步引出结论，最后以总结归纳结，但是在组织英语篇章时，通常把重要的信息放在前面首先说出，以一个主题句开头，直截了当地陈述主题，点出这个段落的中心思想，然后将其分解，逐个论述，因此建议将"这是概率风险评估技术的一部分"最先译出。

译文：As part of the probabilistic risk assessment technique, as reactor and systems designs become solidified, an in-depth event sequence analysis should be needed, including a full exploration of the initiating events that can happen during startup, full-power operation, shutdown, and dormancy periods.

练习
practice

1. 简答题。
(1) 翻译过程有哪些环节？
(2) 理解原文应注意哪些部分？
2. 赏析下列翻译，指出译文的优点。
(1) 为了提高实用性，大多数3DTV供应商并不考虑提供刚才所说的全分辨率和/或传输两个完全独立的频道(同步广播)，而是采取一些折中办法；例如，降低每只眼睛的数据速率，使3DTV节目适合商业级的HDTV频道(如8~10 Mb/s)，使用时间交错或空间压缩—同样，这是一个可行方案，但最终分辨率质量会下降。

To be pragmatic about this issue, most 3DTV providers are not contemplating delivering full resolution as just described and/or the transmission of two fully independent channels (simulcasting), but some compromise; for example, lowering the per eye data rate such that a 3DTV program fits into a commercial-grade HDTV channel (say 8-10 Mb/s), using time interleaving or spatial compression—again, this is doable but comes with the degradation of ultimate resolution quality.

(2) 当右眼视图中的点比左眼视图中的相应点更靠右时，相应的图像点被称为具有正视差或不交叉视差 P。

Corresponding image points are said to have positive or uncrossed parallax P when the point in the right-eye view lies more to the right than the corresponding point in the left-eye view.

(3) 一个矩形可以定义为四个数字，但要定义一个更复杂的形状，就需要一个具有更多参数的更复杂的模型。

A rectangle can be defined by four numbers, but to define a more complicated shape one needs a more complex model with a much larger number of parameters.

(4) 通过将行星系统作为一类天体物体进行研究，从它们在环恒星或环行星盘内的形成到其中可能出现的宜居世界，人们可以补充星周盘、太阳系物体和系外行星之间的"观测空白"，更好地解决有关它们的关键科学问题。

By studying Planetary Systems as a class of astrophysical objects, from their formation inside circumstellar or circumplanetary disks to the possible emergence of habitable worlds within them, one can bridge the "observational gaps" between disks, solar system objects and exoplanets and better address key scientific questions about them.

(5) 结束载人任务推进阶段时需要将反应堆在"日心空间"处理掉。

End of life for the crewed mission propulsion stage would involve disposal of the reactors into "heliocentric space".

3. 在正确理解原文的基础上翻译下列句子。

(1) 它们是由一个主系统的天体(其卫星、星环、气体群、尘埃粒子、等离子体和带电粒子)的一个子集形成的，这些天体主要与一个特定的行星耦合，并因为重力和磁场的作用限制在周围有限的空间内。

(2) 多层粒子阱是一种人工神经网络结构，是一种可用于分类和回归的非参数估计器。

(3) 类海王星是对核心吸积形成巨行星这一观点的关键检验。

(4) 我们对这些卫星的组成和内部动力知之甚少。

第二节　查阅平行文本

平行文本(parallel text)常出现于比较语篇语言学当中，应用语言学家莱茵哈德·哈特曼(Reinhard Hartmann)将平行文本分为三类，即形式上非常一致的译文及原文、形式上不完全一致但功能对等的译文及原文、在同样交际情境中独立产生的两种不同语言的文本。而阿尔布雷希特·纽伯特(Alberecht Neubert)从翻译的角度将平行文本定义为"在大致相同的交际情景中产生的具有相同信息性的双语文本"。朱莉安·豪斯(Juliane House)则将平行文本定义为"产生于不同语言文化环境，但属于相同的体裁和文本类型，并且具有相同功能的文本"。李长栓认为平行文本指并排放在一起、可以逐句对照阅读的原文及其译文，这是对平行文本的狭义解释；广义的平行文本还包括与原文内容相似的译出语资料，主要用于帮助读者深入理解原文。简单地说，平行文本就是与原文内容接近的任何参考资料。

因此，广泛查阅平行文本可以弥补专业知识和语言能力的不足。科技文本旨在传递科技经验、陈述科技事实、回应技术问题。平行文本具有与原文相同的功能性，可以帮助译者理解原文，避免译者忽略原文的预期目的；同时，平行文本体现着文本的体裁规范，由于源语国家和目标语国家之间存在文化、体裁规范的差异，平行文本可以为译者提供目标语国家的体裁规范。

目前获取平行文本的途径颇多，传统方式是查阅特定学科专业领域的报纸、文献等。但随着互联网的发展，丰富的网络资源成为查阅平行文本的主要途径，译者可以在必应、维基百科等进行搜索，同时许多公共科技英语语料库也可供使用，如 WMT、IWSLT、LDC 等。译者查阅平行文本时，可根据关键词进行检索，查找与原文文本意义相同或相近的文本，也可查找功能相同的文本作为参考，但由于平行文本也具有不准确性，译者在查阅平行文本时应参考多个平行文本并仔细甄别。

翻译科技文本时，译者可能缺乏相关领域的专业知识，查阅平行文本有助于译者了解原文本中专业术语的翻译，保证术语准确，提高专业知识储备。同时，由于中英文之间存在差异，查阅平行文本有助于加深译者对原文以及目标语的理解，选择灵活的翻译策略，借鉴表达方法，达到忠实、通顺、准确。

【经典赏析】

原文：XXX，男，1962 年，江苏阜宁县人，毕业于 XXX 大学，分别于 1982、1985、1990 年获工学学士学位、工学硕士学位、博士学位。现任 XXX 大学通信工程学院院长、信息科学研究所所长、博士生导师、中国通信学会会士、电气与电子工程师协会高级会员、中国电子学会高级会员、获陕西省和电子工业部"有突出贡献的专家"称号，享受政府津贴。

平行文本 1：Jiandong Li received the M.S. and Ph.D. degrees from Xidian University, Xi'an, China, in 1985 and 1991, respectively. Since 1985, he has been a Faculty Member with the School of Telecommunications Engineering, Xidian University, where he is currently a Professor. From 2002 to 2003, he was a fellow of the China Institute of Electronics and the China Institute of Communication. From 1993 to 1994, and from 1999 to 2000, he was a member of Personal Communications Networks. (IEEE.org)

平行文本 2：Professor K.T. Chau received his B.Eng. degree with First Class Honours, M.Phil. degree, and Ph.D. degree all in Electrical & Electronic Engineering from the University of Hong Kong. He is a fellow of the IET, and fellow of the HKIE. He is elected fellow of the IEEE for Contributions to Energy Systems for Electric and Hybrid Vehicles. He has served as one of organizing committee members for many international conferences, especially in the area of Electric Vehicles. (Hong Kong University)

译文：XXX，male and born in Funing County, Jiangsu Province in 1962. He graduated from XXX University and received the B.Eng degree, the M.Eng degree and Ph.D degree in 1982, 1985 and 1990 respectively. In 1994, he was exceptionally promoted to professor. Presently he holds the posts of dean the School of Telecommunication Engineering and director of the Research Institute of Information Science at XXX University. He is an advisor for doctoral students, a fellow of the Chinese Institute of Communications, a senior member of IEEE, and a senior member of the Chinese Institute of Electronics. He is granted the title of "Expert with Distinguished Achievements" of both Shaanxi Province and the Ministry of Electronics Industry. He is a recipient of the Government Special Subsidy.

赏析：阅读原文不难发现这是 XXX 大学 XXX 教授的简介。基于此，译者可以在 XXX 大学官方网站进行检索，或在必应等搜索引擎中搜索该教授信息，便

可找到其英文简介，如平行文本 1 来源于 IEEE 官网。此外，译者可以查阅
港澳台地区或国外学校官网，获取同领域教授的英文简介，如平行文本 2
所示。译者可参考平行文本进行翻译。但由于中英文的差异性，原文本是
一个长句，译者阅读平行文本可发现，平行文本并不是一个长句，而是拆
分信息，因此在译文中也需考虑这一因素，采用适当的翻译方法。

【案例解析】

◇ 例 1

原文：不幸的是，经常有低品质的纺织品供应，可能产生疵件，因此拉布和裁
剪时间增加，引发客户和厂家之间产生纠纷。

原译：Unfortunately, low-quality textiles are often supplied, which can lead to
defects, resulting in increased paving and cutting time, causing disputes
between customers and manufacturers.

解析：阅读发现原文是涉及纺织领域的表达，因此出现了"拉布""裁剪""疵
件"等专业表达。在原译中，由于欠缺纺织领域的专业知识，以上术语
翻译出现偏差。比如，查阅释义可知"拉布"是"将卷状原布料按设计
长度拉出并平铺于裁床上"，平行文本中译为 spreading 而非按照字面意
思译为 paving。

译文：Unfortunately, low quality textile materials are often supplied and the
spreading and cutting time therefore increases, with the potential to produce
faulty articles, giving rise to disagreements/disputes between customers and
producers.

◇ 例 2

原文：由于局域表面等离子体共振这一特性众所周知，这些纳米颗粒具有独特
的光电特性。

原译：Because of the well-known property of local surface plasmon resonance, these
nanoparticles have unique photoelectrical properties.

解析：原文出现了一个光学电子领域的专业术语"局域表面等离子共振"，因
此译者可在科技英语词典中进行检索，便可得知该术语被译为 Localized
Surface Plasmon Resonance (LSPR)。随后译者可检索纳米颗粒 LSPR 特
性的文本，得到平行文本作为参考。

译文：Due to well-known Localized Surface Plasmon Resonance (LSPR) characteristics,
these nanoparticles possess unique optoelectrical properties.

◇ 例 3

原文：卷维是指数据的规模。大数据生态系统中的数据大小范围可以从数十 TB
到几 ZB，并且仍在增长。

原译：Convolutional dimension refers to the size of the data. Data sizes in big data
ecosystems can range from tens of terabytes to several zettabytes and are
still growing.

解析：原文翻译难点在于"卷维"这一术语，检索关键词"大数据"或"big
data"，可得知通常从四个维度衡量大数据，即 volume、velocity、variety、
accuracy。其中，volume 就指代数据的规模大小，可知 volume dimension
即为"大数据的卷维"。原译中"卷维"的翻译有误。

译文：The volume dimension refers to the amount of data. The data size in a big
data ecosystem can range from dozens of terabytes to a few zettabytes and is
still growing.

◇ 例 4

原文：人们对于病变类型、空间分布和程度与创伤性脑损伤之后发生的急性和
慢性后遗症之间的关联仍知之甚少。

原译：People at present know little about the relation between lesion type, spatial
distribution, and extent with acute and chronic sequelae following traumatic
brain injury.

解析：原句属于医学领域，难点在于对"病变""创伤性脑损伤"等术语的翻
译。译者可以查找术语库，即可发现医学术语的标准翻译，原文中术语
翻译准确。但是，原文是以"人们"作为主语，查阅相关的平行文本，
可知这是由于中英语言差异造成，英文文本中惯用被动句式，因此译者
在翻译时应考虑转换句式。

译文：The associations of different lesion types, their spatial distribution and extent
with acute and chronic sequelae after Traumatic Brain Injury (TBI) are still
poorly understood.

练习
practice

1. 简答题。
(1) 平行文本是什么？

(2) 哪些语料库可供查阅平行文本？

2. 赏析下列翻译，指出译文的优点。

(1) 大数据分析是指对从各种来源收集的大量数据进行分析的策略，包括社交网络、交易记录、视频、数字图像和各种传感器。

Big data analysis refers to the strategy of analyzing large volumes of data gathered from a wide variety of sources, including social networks, transaction records, videos, digital images, and different kinds of sensors.

(2) 图灵机是一种简单的抽象计算设备，旨在研究在顺序计算机上可以计算的范围。

A Turing machine is a simple abstract computational device intended to investigate the extent of what can be computed on a sequential computer.

(3) 由于天线安装空间有限，随着天线数量的增加，天线单元的尺寸和间距必须减小。

Due to the limited space for antenna installation, with the increase of the number of antennas, the size of antenna elements and the spacing between them must be reduced.

(4) 中国成为第三个成功从太空中回收其有效载荷的国家。

China became the third nation to successfully recover its pay-load from space.

(5) 航天器寿命终结时的脱轨方法包括化学火箭推进、电力推进、风帆、系绳等。

Methods for end-of-life de-orbiting of spacecraft include chemical rocket propulsion, electric propulsion, sails, tethers, etc.

3. 查找平行文本，翻译下列句子。

(1) 卫星将收集、研磨并分析近 70 个来自火星的岩石和土壤标本。

(2) 内质网是一种细胞器，蛋白质和脂质在其中进行合成和修饰，它还充当钙库。

(3) 在受到自然灾害影响的地区，城区雨水蓄积有双重的功效：补充自来水供应；通过收集、利用有效减少多余的城市径流来改善城区水文状况。

(4) 显卡是出了名的耗电器件，产生的热量可以让一间房子变暖。

(5) 如今，考虑到最近一波的网络攻击，机密性、完整性、可用性、真实性、不可更改性、可靠性、隐私性和授权访问控制等安全需求在车辆环境中也变得至关重要。

第三节　词汇翻译

词汇是构成文本的基础元素，科技词汇在科学技术的带动下不断发展变化。在进行科技文本汉英翻译时，要秉持准确、专业、精炼的原则，使译文符合英语的惯用表达法，摆脱汉语腔调，妥善处理词类转换、词义引申、科技术语翻译等方面。

1. 词类转换

汉语和英语分属于两个不同语系，因此存在部分词汇无对等词义的情况，翻译时要从汉语的基本含义推敲，选择英语中相应的词语进行确切表达。为了使译文更符合英语语法和惯用法，有必要改变原文本中部分单词的词类，用准确、顺畅的语言传递出原文信息，因此在汉英翻译过程中，词类转换现象十分常见，动词、名词、形容词等词类在翻译过程中可以相互转换。例如在科技文中"一经点火"这一表述可以译为 at ignition，就是采用了词类转换方法将动词译作名词，更符合科技文本简洁严谨的特点。

2. 词义引申

译者应从单词的基本意义出发，结合上下文考虑全句涵义，选择适当的英文表达来传递原文意义，而非逐词硬译，这就是词义引申法。

常用的词义引申可归纳为三种方法：

(1) 近似法，即根据汉语基本意义，不受词典所约束，力求选用近似的英语来表达，即使形式上超出词典所注释的词义范围，但是能够确切表达出原文含义。

(2) 第二种为上下文体会法，即由上下文语境体会出某个单词所指的人或物，避免逐词译为英文中不可理解的词语。

(3) 第三种是引用英文惯用语，包括口头或书面常用词组或谚语，但此方法在科技文本的翻译中较为少见。

3. 科技术语翻译

随着现代科学技术飞速发展和我国对外科技交流日益频繁，在翻译过程中如何妥善处理科技语并提高翻译质量已成为一项重要研究课题。科技词汇主要具有以下几种特点：

(1) 专业术语繁杂且词义专一，基本上无一词多义现象，如 chromosome(染色体)、anatomy(解剖学)、photosynthesis(光合作用)等。

　　(2) 非技术词汇专业化，除少数用途极其狭窄的专业词汇外，相当多的英语单词除常见意义外，都具有在其他领域的专业意义。

　　(3) 大量使用缩略词，包括首字母缩略词和缩略部分字母构成新词两种，如 GPS (Global Positioning System)、nanolaser (nanometre + laser)等。

【经典赏析】

原文：本周一，我国第 2000 万辆新能源汽车在广东省广州市下线，这是中国新能源汽车发展的历史性时刻。这标志着中国新能源汽车迈入规模化、全球化的高质量发展的新阶段。

译文：China's 20 millionth New Energy Vehicle (NEV) rolled off the production line on Monday in Guangzhou, capital of Guangdong Province, creating a remarkable milestone for the country's NEV sector. With this record, the NEV sector has entered a new stage of large-scale, globalized and high-quality development.

赏析："新能源汽车"翻译为缩略形式 NEV，"历史性时刻"引申为 milestone。"这标志着"本为主谓结构，在译文中处理为状语 with this record。

【案例解析】

◇ 例 1

原文：设备中传感器数量增加必然会产生更高的数据吞吐量，这对管理以及加工处理海量的感知信息造成了严峻挑战。

原译：If the number of sensors in devices increases, it will inherently generate higher data throughput, which poses a serious challenge in managing and processing the tremendous amount of sensory information.

解析：原文中"数量增加"一词是动词，在科技翻译中采用词类转换方式，将动词处理为名词更符合英文表达习惯，同时使译文更加简练且逻辑更为清晰，也避免条件状语从句导致句子冗长。

译文：An increase in the number of sensors in devices will inherently generate higher data throughput, which poses a serious challenge in managing and processing the tremendous amount of sensory/perceptual information.

◇ 例 2

原文：CNN 分类结果专门用于识别患有黑叶斑病的香蕉叶片，每一张图像的轮廓都是彩色的。

原译：The results of CNN classification specifically identify whether the banana leaves suffer from Black Sigatoka disease, and the outline of each image is colorful.

解析：原文中"患有"是动词，用词类转换法，将动词处理为介词 with，避免使用从句导致句子结构复杂；"每一张图像的轮廓都是彩色的"采用词义引申的方法，从单词的基本意义出发，结合上下文考虑全句涵义，巧妙处理为英文固定词组搭配，而非逐词硬译。

译文：The results of CNN classification specifically identify banana leaves with the Black Sigatoka disease, and each image is outlined in color.

◇ 例3

原文：这种简易的传感器系统很容易在某种气体环境中失效，因为未知的气体种类可能会与二氧化碳测量值产生串扰。

原译：However, such a simple sensor system may easily fail in a gaseous environment where unknown gas species and the CO_2 measurements will produce series interference.

解析："串扰"一词正确译法为 crosstalk，体现了科技词汇中非技术词汇专业化的特点，crosstalk 一词除常见的"相声"之义外，还有"串扰"这一专业意义。同时，本句翻译采用词类转换法，将名词"串扰"处理为动词 crosstalk，使整句结构更紧凑，句意更明确。

译文：However, such a simple sensor system may fail easily in a gaseous environment where unknown gas species may crosstalk with the CO_2 measurements.

◇ 例4

原文：该模型将图像分为四类，包括正常组织、良性病变、原位癌和浸润性癌。

原译：The model classifies the images into four classes, including normal structure, benign disease, in situ carcinoma, and invasive carcinoma.

解析：翻译该句时，须注意专业术语的准确性，翻译专业术语必须经过严谨的查证。生物学中的"组织"，应译为 tissue；"病变"应译为 lesion 而非 disease。

译文：The model classifies the images into four classes, including normal tissue, benign lesion, in situ carcinoma, and invasive carcinoma.

练习
practice

翻译下列句子，注意画线部分词汇翻译的方法。

(1) ML 算法具有<u>通用性</u>和<u>鲁棒性</u>，因此能够适应几乎任何对任意数据集有基本要求的应用程序。

(2) 与非 NN 算法相比，NN 算法在特征学习和特征提取方面<u>具有很高的效率</u>，需要的人工输入更少。

(3) 输入刺激数据传递到整个 NN，<u>以数学的方式</u>激活每个神经元节点。

(4) 生乳中添加了<u>福尔马林、过氧化氢、次氯酸钠</u>等化学物质来延长保质期。

(5) 图 3a-3 显示出，与 MLR 相比，转换后的 PLS 分量呈线性集群。

(6) 可以在二维或三维平面的 PCA 双标图上将其<u>直观地呈现出来</u>。

(7) 未知混合物的气体测量结果<u>用红色集群表示</u>。

(8) RNN 与 FNN <u>具有相同的属性</u>。

(9) 物理和化学传感系统中的传感器<u>需要与目标环境相互作用</u>，这会改变传感器本身的化学特性或物理特性，从而产生可测量的信号。

(10) 将一维原始传感器数据堆叠在一起<u>形成行</u>，生成二维信号图像。

第四节　句子翻译

句子翻译是翻译的基本功，是翻译实践训练中必不可少的重要内容。但相当一部分译者不能正确认识汉英两种语言的差异，无法掌握句子翻译的技巧和方法并正确使用。因此造成译文混乱无序，逻辑不通，常常带有严重的翻译腔，无法正确传达原文意思。本节将通过介绍科技文体汉译英中句子翻译的技巧，给译者提供翻译实践方面的帮助。

1. 汉英句子结构差异

汉英两种语言分属不同语系，行文构句时在语法、用词、组织结构、行文顺序等方面都存在差异。要熟练掌握汉英翻译，正确认识两种语言的差异是关键的一环。

汉语和英语句子中最重要的语言区别是汉语意合与英语形合的区别。汉语属意合，通常是以时间顺序层层递进；而英语多用形合方式组合句子，借助语言形式手段(词汇、形态)来链接词语或句子。英语句子是一种以动词为中心的空间结构体，也就是以主谓结构为主干，以谓语动词为中心，并通过大量关系词、连词、介词把句子的各个成分分层搭架而成，从而呈现出由中心向外扩展的空间形式(冯庆华，2002)。因此，英语中连词使用较多。

【经典赏析】

原文：自 1992 年我国首个新能源汽车项目正式立项以来，中国在发展新能源汽车领域取得了令人瞩目的成就，目前新能源汽车的产量和销量已连续八年位居全球第一。

译文：Since China registered its first NEV project in 1992, the country's achievements in developing its NEV sector have been remarkable, with its current production and sales volumes for NEVs both ranking the first worldwide for eight years in a row.

赏析：原文最后两句话之间仅以逗号隔开，译文在最后部分使用了 with 结构，符合科技英语表达习惯。

【案例解析】

原文：氢能对电力系统的零碳转型至关重要，它适合应用在大型风电、光伏等可再生能源配套系统上，实现跨季节和跨地区动态储能。

原译：Hydrogen is crucial to the net-zero transition of the power sector, it matches well with large-scale renewable energy such as wind and solar and achieves dynamic energy storage across seasons and regions.

解析：译文第一小句和第二小句之间无连接词，不符合英语规范，第二小句和第三小句之间的逻辑关系不清晰。可增添连词或通过语法手段使逻辑关系显性化。

译文：Hydrogen is crucial to the net-zero transition of the power sector in that it matches well with large-scale renewable energy such as wind and solar to achieve dynamic energy storage across seasons and regions.

2. 汉语中长句的译法

长句是字数较多、结构复杂、含有多层意思的句子。汉英翻译过程中处理长句最常用的方法是分解句子，先正确理顺原文意思以及强调的重点，再进行翻译。

【经典赏析】

原文：据中国氢能联盟估计，到 2030 年，中国氢能市场规模中绿氢的占比将从 2019 年的 1%提高到 10%，市场规模将增长近 30 倍，这将需要每年至少生产 500 万吨氢。

译文：Green hydrogen will take up more in the energy mix, from 1 percent in 2019 to 10 percent by 2030, the China Hydrogen Alliance estimated, adding the market scale will have increased nearly 30 times by then, which would require the production of at least 5 million tons of hydrogen per year.

赏析：从原文可以判断出，本句第一个重点信息为"中国氢能市场规模中绿氢的占比将提高"，翻译为句子主干，另一同等重要的信息"市场规模将增长近 30 倍"处理为分词短语，adding 意为"并补充说"，表示进一步解释说明，其余部分均处理为状语或定语从句。译文重点突出，符合英文表达习惯。

【案例解析】

原文：随着北京市高级别自动驾驶示范区工作办公室宣布，正式开放智能网联乘用车"车内无人"商业化试点，扩大服务区域，将有更多无人驾驶出租车在北京投入运营。

原译：As the Office of the Beijing High-Level Autonomous Driving Demonstration Zone has announced that the commercial operation of autonomous vehicles will be started and the serviceable area will be expanded, more driverless taxis will be put into operation in Beijing.

解析：本句为典型长句，从原文可判断出主干为末尾部分的"将有更多无人驾驶出租车在北京投入运营"，在翻译时可置于句首，使重点信心突出，再综合考虑句型结构、简洁性、准确性等多个因素。

译文：More driverless taxis will soon be available for hire in the capital city as the Office of the Beijing High-Level Autonomous Driving Demonstration Zone has decided to launch the commercial operation of autonomous vehicles and expand their serviceable area.

3. 汉语无主句的译法

汉语中只有谓语而没有主语的句子称为无主句。所谓无主句，并不是真正意义上的主语不存在，而有可能是主语不明显、主语不重要、不想突出主语、表达习惯等多方面原因导致的主语省略。因此，在汉译英过程中无主句翻译往往需要补充主语、使用被动句或改变句型。

【经典赏析】

原文：今年 3 月，在进行了一系列道路测试后，无人驾驶出租业务在北京试运
行。上述测试在安全监督员的指导下进行。

译文：Self-driving taxis were rolled out on a trial basis in the capital in March
following a series of road tests, which were conducted with safety
supervisors on board.

赏析："运行"和"进行"均无主语，其所在的两句话均译为被动语态。

【案例解析】

◇ 例 1

原文：必须仔细观察病人是否有迟发性呼吸抑制的体征。

原译：Doctors must carefully observe the patient for signs of delayed respiratory
depression.

解析：原文强调专业领域规范操作，施事者并不重要，汉译英时可采用被动句。

译文：The patient must be carefully observed for signs of delayed respiratory
depression.

◇ 例 2

原文：已经证明，用最精密的方法加工的金属表面也不是真正平的。

原译：We have proved that even the most carefully finished metal surface is not
truly flat.

解析：原文突出事件，不必要强调主语，汉译英时可采用主语从句句型"it is
proved that…"。

译文：It has been proved that even the most carefully finished metal surface is not
truly flat.

4．汉语省略句的译法

省略句即缺少某一个句子成分(主语、谓语、宾语、定语、状语、补语)的
句子。汉译英过程中，省略句的处理方式一般是补充或者省略某一成分。

【经典赏析】

原文：中国科技巨头百度和自动驾驶初创公司小马智行获得了许可，在北京亦
庄经开区 60 km^2 范围内划定区域开展全无人自动驾驶运营示范。

译文：Chinese tech giant Baidu and autonomous mobility startup Pony.ai won permits
to operate driverless taxis through their ride-hailing apps within a 60-square-

kilometer area in Yizhuang, located in the southern suburbs of Beijing.

赏析：原文省略了中国读者熟悉的信息，而译文补充信息 through their ride-hailing apps(通过叫车应用程序)，说明如何开展无人自动驾驶，并对"亦庄"进行补充说明，使国外读者更了解我国无人驾驶试运营区域。

【案例解析】

◇ 例1

原文：虽然能量不能被创造出来也不能被消灭，但可以从一种形式转变成另一种形式。

原译：Energy can be neither created nor destroyed, but can be transformed from one form to another.

解析：原文第二个分句省略了主语，译文进行补充，并使用主从复合句。

译文：Though energy can be neither created nor destroyed, it can be transformed from one form to another.

◇ 例2

原文：根据 1950 年以来的雨雪测量结果以及电脑模拟未来气候的数据，科学家估算出这样一个结果：全球温度每升高 1 度，高海拔地区发生特大暴雨的概率就会增加 8.3% (每升高 1℃概率增加 15%)。

原译：Using rain and snow measurements since 1950 and computer simulations for future climate, scientists calculated that for every degree the world warms, extreme rainfall at higher elevation increases by 8.3% (15% for every degree Celsius).

解析：原文的"1 度"信息不够精确，译文补充信息，将其明确为"华氏温度" Fahrenheit。

译文：Using rain and snow measurements since 1950 and computer simulations for future climate, scientists calculated that for every degree Fahrenheit the world warms, extreme rainfall at higher elevation increases by 8.3% (15% for every degree Celsius).

练习 practice

翻译下列句子。

(1) 在过去的二十年里，我们见证了各种革命性移动设备的出现，例如智能手机和可穿戴设备，这些设备带来了移动计算的繁荣，使人们可以随时随地进行计算和通信。

(2) 电子媒体使用的增多以及睡眠质量的下降、睡眠时间的缩短很可能对精神健康起到了核心作用。尽管青少年面对着城市化带来的变化以及来自学校的压力，这些生活中的变化对精神健康也起到了一定的作用。

(3) 数字时代带来的机遇和复杂性可能会使行业和市场难以承受，因为它们在每笔交易中都面临着大量的潜在信息。

(4) 使用智能学习方法的总体目标是训练机器在与人类相同的和不同情况下都能进行智能思考和决策。

(5) 数字时代数据生产的快速增长引入了大数据的概念，大数据以其显著的容量、多样性、准确性、速度和高价值来定义，但同时也给分析带来了挑战，要求组织部署新的分析方法和工具，以克服不同数据类型的规模和复杂性。

第五节   修 改 译 文

译者通过译前准备和译中翻译环节，在原文和译文间进行基本语义转换。一般情况下，翻译初稿更多集中在对语言层面的理解，而科技类文本的内容远超出单纯的语言层面，涉及大量科技知识、技术数据、公式图表等。因此，科技类文本初稿翻译中难免出现语言理解、标点符号、数据图表、计量单位、内容信息等方面的遗漏或差错，这就需要译者对照原文反复阅读、修改。

词汇方面，科技类文本专业术语多，科学性很强。译者通过查阅平行文本、请教专业人士后确定专业术语英译表达，建立术语表，此后应严格按照术语表进行翻译，保证术语一致。术语统一不仅是个人翻译时要注意的问题，对多人合译同一文件也有重要意义。如果各译员之间术语不统一，不仅要浪费时间与精力逐一修改，还会因译法混乱造成理解偏差。

句法方面，科技类文本句子结构复杂，长难句多，且句间多隐藏逻辑关系。翻译过程基本以句子为单位，因此要特别重视句与句之间的衔接与连贯。对于一些句间隐性的逻辑关系则需要重新梳理句子结构，借助代词或连接词适当显化，避免逐词翻译。

文体方面，科技类文本主要以传递信息为主，严肃正式，具有客观性。因此翻译过程中既要保证译文与原文风格一致，还要做到信息最大化的同时保证译文整体风格一致。

　　审校修改 对译文进行审校修改的目的是发现并改正译文中可能出现的错误。通常情况下，审校译文中的错误不应只停留在标点符号或字词拼写上，更重要的是修正违背原文内容的信息。

　　审校通常分为自我校对和他人校对。译者在完成翻译任务后，首先需要对译文进行自我校对。在反复阅读的基础上，首先对标点符号、单词拼写、语法等进行核查；其次检查专业术语前后是否一致、有无过译漏译情况、句间逻辑是否清晰连贯；最后从语篇角度，检查译文整体风格是否与原文一致，语言是否通俗易懂、简洁明晰。自我校对后，需要寻求领域内专业人士或英语母语者在脱离原文的基础上对译文进行研读和润色，重点核查译文信息是否有误、英文表达是否流畅地道、符合规范。总而言之，修改译文时要把握住两大原则：保证专业性、提升可读性。

【经典赏析】

原文：任何一种保护元件都具有与其钳位特性相关联的固有电阻值，如果串联电阻高，元件上就会出现高压，降低对集成电路的有效保护；如果串联电阻低，保护元件上产生的电压会降低，集成电路暴露水平也会降低。这种较低的动态电阻允许更多的浪涌电流从集成电路中流出并接地。

译文：Any protection component has an intrinsic resistance value associated with its clamping characteristic. If its series resistance is high, a high voltage develops across the component, providing less effective protection for the IC. If its resistance is low, the voltage developed across the protection component is reduced, and thus the exposure level to the IC is decreased. This lower dynamic resistance allows more of the surge current to be routed away from the IC and into ground.

赏析：译文虽与原文语序一致、内容相同，但根据科技英语的特点将长句进行了拆分，按原文表达的意义和逻辑关系选择使用代词 its 和连接词 thus 等，保持句内、句间衔接连贯。这符合科技英语表达习惯，不仅使句子通顺连贯，语篇的整体内容也衔接得更加紧密。

【案例解析】

◇ 例 1

原文：理想电压源和电流源都是有源元件，而电阻和导体都是无源元件。

原译：Both ideal voltage source and ideal current source are active elements, while both resistor and conductor are passive elements.

解析："有源元件"和"无源元件"是电子、通信领域的专业术语，需要译者在翻译前期进行查阅。active 和 passive 的本义为"积极的"和"消极的"，在此领域意为"有源"和"无源"。ideal voltage source and ideal current source 可合并处理为 ideal voltage and current sources，这样更符合科技英语语言简洁的特征。此外，ideal voltage source and ideal current source 和 resistor and conductor 均应使用复数形式。

译文：Ideal voltage and current sources are active circuit elements, while resistors and conductors are passive elements.

◇ 例2

原文：卷积神经网络与密集神经网络相比非常不同，因为它将多个输入连接到单个输出，而不是将每个输入连接到每个输出。

原译：Convolutional neural network is very different from dense neural network, which connect multiple inputs to a single output instead of connecting every input to every output.

解析：原文中"因为它将多个输入连接到单个输出，而不是……"一句中，"它"指的是"卷积神经网络"，而原译中 which 作为关系代词引导非限定性定语从句，修饰的是先行词 dense neural network，这与原文意思相悖。因此，为避免关系代词指代不清，译文选择尊重原文语序，用原因状语从句 because it...表达，其中 it 指代的是 convolutional neural network。另外，还需关注细节，比如主谓一致及冠词使用等。

译文：The convolutional neural network is very different from the dense neural network, because it connects multiple inputs to a single output, as opposed to every input to every output.

◇ 例3

原文：它的优势在于，它不是作为根目录或全局目录安装在系统下，而是安装在主目录中。我们可以将其添加到现有系统中，而不必担心会破坏可能依赖 Python 的系统构件。

原译：Its advantage is that it is not installed as a root or global directory on the system, but rather in the main directory. We can add it on to an existing system without worrying about breaking system components that might rely on Python.

解析：在计算机领域，"根目录""全局目录"和"主目录"分别表达为 root directory、global directory 和 home directory。这两句话前后暗含逻辑关系,"因为"有第一句话这样的优点,"所以"有了第二句话中的结果。在翻译过程中我们用表示因果关系的连词 therefore 显化句间逻辑,使句子更为连贯。此外,科技文本多用被动语态表示客观性,因此第二句话可处理为被动式 it can be added on to...。

译文:It has the advantage of being installed, not as a root or global directory underneath the system, but in the home directory. Therefore, it can be added on to an existing system without worrying about breaking system components that might rely on Python.

◇ **例 4**

原文:这就是密集神经网络的定义:所有输入和所有激活之间完全连接,所有激活和所有输出之间完全连接。

原译:That's what defines a dense neural network: the full connection between all inputs and all activations, and the full connection between all activations and all outputs.

解析:connection 强调的是人际、事件之间的"关系";connectivity 强调的是科学技术、网络方面的"连接",在本句中宜用 connectivity 表示二者之间的连接性。此外,由 and 连接的句子中,为避免重复常省略共同的成分。此句中,the full connectivity between 可省略,以保证句子的简洁性。原译中 and 较多,可以选择 as well as 作连词,连接前后并列的两句,使句间关系更为明晰。

译文:That is what defines a dense neural network: the full connectivity between all inputs and all activations, as well as between all activations and all outputs.

练习 practice

1. 简答题。

(1) 修改译文的重点在哪些方面?

(2) 如何对译文进行审校?

2. 评析下列翻译。

(1) 由此可以推测,与任务导向型文化相比,关系导向型文化具有更强的并行信息处理能力,其交互频率和交互速度以及信息密度和信息传递速度都更高。

Therefore, it can be supposed that interaction frequency and interaction speed as well as information density and information transfer speed is higher for relationship-oriented cultures, which exhibit rather more parallel information processing in contrast to task-oriented cultures.

(2) 行动链由实现目标的步骤表示，例如问候、教育、合作合同、合资、发行股票、创业、决断乃至打高尔夫球。

Action chains are represented by steps to reach an objective, such as greetings, education, contracts of cooperation, joint ventures, stock emission, entrepreneurship, judgment or even playing golf.

(3) 池化操作的作用是减小图像尺寸并聚焦于最强的值。

What the pooling does is to reduce the size of the image and focus on the strongest values.

(4) 特别是在研究 AI 后的 80 年中，由于当代技术的限制出现了繁荣和低谷，如难以处理非线性问题、特征提取过程中缺乏人类专业知识等。

In particular, within the 80 years of AI research, there have been periods of upsurge and downturn, due to the restrictions of the contemporary technology, e.g., difficulty in handling nonlinear problems, the necessity for human expertise during feature extraction, etc.

(5) 通常情况下，样本总是多维数据数组中的第一个维度。在这里，我们有多个样本，因为机器学习的基本工作原理是通过查看大量不同样本中的大量不同数据点学习一个函数，并在此基础上预测结果。

By convention, samples are always the first dimension in the multidimensional array of data. Here, we have multiple samples because machine learning fundamentally works by looking at a wide array of different data points across a wide array of different samples and then learning a function to predict outcomes based on that.

3. 翻译下列句子。

(1) 实际上，在机器学习中，将数据分割成片段是一种常见的方式。

(2) 最基本的张量是一阶张量，在编程语言中称为数组。它是一个由单个数字组成的有序序列。

(3) 这正是机器学习算法中数值稳定性所需要的。

(4) 训练一个神经网络通常需要提供大量的数据，我们称之为数据集。数据集一般可分为三类，即训练集，验证集和测试集。

(5) 光纤系统的主要损耗通常发生在发射器和接收器的电光转换过程中，

实际的电缆损耗大约是同轴电缆的 1/3。

第六节　译　后　编　辑

译后文档编辑是提高文本翻译质量的重要一环。无论是机器翻译还是人工翻译，都需进行译后编辑。译后编辑是基于译者对翻译整体内容的把握，针对目标语读者的阅读习惯和要求修改、完善译文。译后编辑涉及词汇、句子以及语篇三个层次。

词汇方面，最基本的是要做到专业词汇的准确翻译，避免术语翻译错误。由于英文词汇语义和词性的多样性，译者在翻译过程中容易出错。在进行译后编辑时，译者要着重关注此类词汇，避免词性判断出错。与此同时，由于中英文标点符号的使用存在差异，要仔细检查是否存在标点符号的表征形式(如全角/半角)等方面的错误，遵循目标语读者的标点符号表征习惯。专有名词、函数、运算符号以及拓展名等可保留原样。

句子方面，首先要厘清逻辑关系。英文是低语境文化语言中极具代表性的一种。英文直接明了，注重句与句之间的逻辑关系，容易出现长句。汉语短句较多，汉译英过程中要补充其中隐藏的逻辑关系，做好句与句、段与段之间的承接。同时要注意调整句子位置，避免原文与译文句子长度比例失调。其次，特殊句式的翻译要把握好忠实原则，例如倒装句、否定句、强调句等。在翻译过程中，要凝练精准，删减冗余。中文的四字短语也是一大难点，译者要努力做到"形似"和"神似"兼备。

语篇方面，首先要保持整体风格的一致性，符合主题表达风格。科技文本多为严肃正式的文本，全文应该保持一致的文风。其次，在汉译英过程中，由于译者缺乏背景知识等原因，可能会出现过译和欠译的情况。科技文本翻译最重要的是要做到忠实、准确地传达相关知识，因此译者要仔细查错并且查阅相关平行文本或者请教师长，修改完善译本。不同语言有着独具特色的表达，在翻译过程中做到忠实、通顺之外，还应该追求"美"，要将文化内涵和文化精髓蕴含于译文当中，既要传达出本国语言的优秀文化，又要符合目标语读者的阅读和使用习惯，从而实现"美"的期许。

译后编辑是提高译者翻译水平，实现译本高质量完成的重要手段。译后编辑能够发现翻译过程中存在的不足，提出相应的改善措施，培养译者的全局观。既要找出形式、格式、术语等微观方面的问题，也要找出逻辑、结构、行文、文化等宏观方面的问题，从而实现翻译目的。

【经典赏析】

原文：天问二号任务已正式获得国家批准立项，计划通过一次发射，实现从近地小行星 2016HO3 采样返回地球，之后前往主带彗星 311P 开展伴飞探测。

译文：The scheduled Tianwen-2 mission will sample an asteroid called 2016 HO3, return to Earth, and then head out again to a main-belt comet 311P. The mission has obtained official approval.

赏析：译文首先调整了句子结构，将原文的第一句话单独列出来放到译文最后一句，先阐释天问二号的任务的具体内容，最后说明天问二号正式获得国家批准立项，体现了翻译过程中可以适当调整句与句之间的逻辑关系。

【案例解析】

原文：我们不用那种开关来控制电路。

原译：We do not use those switches to control the circuit.

解析：本句翻译中注意中英文逻辑转换，先表述出语义重点，再使用介词将其他部分连接起来。"控制电路"是主要动作，"使用开关"可以用 with 表达出来。

译文：We do not control the circuit with that kind of switch.

练习 practice

1. 简答题。

(1) 译后编辑主要从哪些方面进行？

(2) 常见的译后编辑需要修正的错误类型有哪些？(至少列出五个)

2. 翻译下列句子。

(1) 原子中心为原子核，电子绕核运动。

(2) 在测量远小于 1 s 的时间间隔时，我们用十进制。

(3) 总体而言，两项研究都表明，将车窗摇下来对燃油效率的负面影响要大于使用车辆的空调。

(4) 4 月 24 日，第四届联合国世界数据论坛在浙江杭州举办。本届论坛有 100 多个国家和地区的 1000 余名代表线下参会，超过 8000 人线上参会。

(5) 4 月 16 日，中国航空工业历史博物馆开馆活动在北京举行。这是国内唯一一座完整反映中国航空工业百年发展历程的主题展馆。

第六章 科技翻译方法

　　方法是指解决思想、说话、行动等问题的门路、程序，是为了达到某种目的而采取的途径、步骤、手段等。方法的英文表述 method 的定义也与此类似：a particular way of doing something (*Collins*)。另外，采用什么方法，不是任意的，而是基于事先确定的一定的原则或方案。基于以上对方法的认识，翻译方法可以定义为是翻译活动中，基于某种翻译策略，为达到特定的翻译目的所采取的特定的途径、步骤、手段。必须指出的是，翻译方法体现的是一种翻译中概括性的处理方式，而非具体的、局部的处理方法。翻译方法是比翻译策略更具体而又比翻译技巧更宏观的一个概念。基于两大翻译策略——异化与归化，常见的翻译方法可相应分为两大类。异化策略下的翻译方法包括：① 零翻译(zero translation)；② 音译(transliteration)；③ 逐词翻译(word for word translation)；④ 直译(literal translation)。归化策略下的翻译方法包括：① 意译(liberal translation)；② 仿译(imitation)；③ 改译(variation translation)；④ 创译(recreation)。以上所列出的八大翻译方法，除了零翻译、音译和逐词翻译外，其他翻译方法都需要在具体的翻译实践中使用翻译技巧。翻译技巧种类繁多，一般包括五种常见的翻译技巧：增译、减译、分译、合译、转译。鉴于翻译策略过于宏观，翻译技巧又太过具体，本章将以科技文本为学习素材，主要介绍三种科技翻译方法。

　　在学习科技翻译方法前，首先要了解目标语——科技英语的主要特征。科技英语是指与科学、技术领域相关的英语，其主要特点如下：

　　(1) 术语丰富。科技英语中包含大量的科技术语，这些术语通常是由拉丁文、希腊文和英文组成的。

　　(2) 精确性高。科技英语中的术语通常具有非常明确的定义，可以精确描述某一具体概念。

　　(3) 标准化。科技英语中的术语通常是标准化的，即在不同的科技领域和语境中都保持统一的定义和使用方法。

　　(4) 语法复杂。科技英语语法结构复杂，包含大量的被动语态、复杂的从句和形容词从句等。

　　(5) 口语与书面语的差异。科技英语在口语和书面语之间的差异很大，通常书面语更加正式、精确。

(6) 全球通用。科技英语是全球通用的语言，在技术领域使用非常广泛，因此成为交流合作的共同语言。

基于科技文本语言平实、语法结构严谨，用词准确的特点，本章将重点介绍逐词翻译、直译和意译这三种翻译方法，也是科技文本翻译中最常使用的三种翻译方法。

第一节 逐词翻译法

逐词翻译是指将原文逐字逐句地翻译成目标语言。在逐词翻译过程中，译者会尽量保留原文的语法、词汇和结构等特征以保证翻译的准确性和完整性。逐词翻译通常适用于技术、法律、科学等领域的文本翻译，而对于文学、艺术等文本则不太适用，因为这些文本涉及更多的文化、情感和艺术表达等方面的内容。

基于科技文本的特点，逐词翻译法是科技翻译中比较常见的方法。简单来讲，逐词翻译就是在翻译过程当中主要采用一一对应的方法。例如，在科技概念的翻译过程中，主要通过字面意思和语法结构进行逐词翻译，以帮助读者更好地理解和学习技术术语。这种从字面来进行逐词翻译的方法在名词概念和定义解释的翻译中具有较大的优势。在信息时代，计算机等方面的技术已经获得了很大水平的提升，出现了很多代表最新科技的技术与专业术语。其中，比较典型的有大数据(big data)、云计算(cloud computing)、网络黑客(network hackers)、人工智能(artificial intelligence)等，这类词都是逐词翻译出来的。但是，由于逐词翻译方法过于机械，其在语言的灵活性等方面存在着一定的局限性，所以在进行科技文本翻译时很难完全对全句进行逐词翻译，常见的应用场景是，句型的基本结构使用逐词翻译，个别修饰语则根据目标语的特点进行微调。

【经典赏析】

原文：虚拟现实(VR)技术允许用户体验一个模拟环境，就像他们实际上在那个环境中出现一样。它已经被广泛应用于游戏、教育、培训等领域。VR技术通常需要专门的硬件，如头戴式显示器或手套，以增强沉浸式体验，以及创建虚拟环境的软件。

译文：Virtual Reality (VR) technology allows users to experience a simulated environment as if they were actually present in that environment. It has been widely used in gaming, education, training, and other fields. VR technology typically requires specialized hardware, such as a headset or glove, to enhance the immersive experience, and software to create the virtual environment.

赏析：本句在翻译时采用全句逐词翻译方法，完全按照汉语的文字顺序进行逐词翻译并形成了语法正确的英语翻译。这里需要说明的是，逐词翻译并不是完全的字面翻译，有时需要调整个别单词的位置或语序以符合目标语的表达习惯。例如，"模拟"被翻译为 simulated，而不是 imitated，因为 imitated 带有个人主观色彩，而 simulated 则有"仿真""模拟"的意思，更适合用于科技文本中。

【案例解析】

◇ 例 1

原文：在极低温度时，比如在绝对零度时，有些导体失去了最后一点点电阻，这一现象被称为超导性。

原译：The phenomenon that at extremely low temperatures, for example at absolute zero, some conductors lose the last vestige of resistance is referred to as superconductivity.

解析：原译将"现象"phenomenon 作为句子的主语，在其后添加同位语从句把汉语前面的主句翻译出来，这使得英语句子的主语太长，显得头重脚轻，失去了英语句子的平衡感。鉴于此，可以采用逐词翻译法将全句译出，既能保证句子的结构均衡，又可体现名词短语做同位语的语法功能。

译文：At extremely low temperatures, for example at absolute zero, some conductors lose the last vestige of resistance, a phenomenon referred to as superconductivity.

◇ 例 2

原文：对谱线进行仔细的研究就会发现，它们中有许多实际上是由紧靠在一起的两根或多根谱线组成的，这一点波尔理论是无法解释的。

原译：If we give a careful study of spectral lines, we will find out that many of them actually consist of two or more close lines. And Bohr theory cannot account for this phenomenon.

解析：原译使用 if 从句，将"对谱线进行仔细研究"处理成了条件从句，这样翻译使得句子前后结构衔接不够紧凑，并且以 we 作主语使得表达客观事实的句子具有主观化色彩。同时最后一个分句被单独翻译出来，从意义到结构表达与前句在衔接上都欠紧密。鉴于此，本句可采用逐词翻译方法，将 careful study 直接作为句子主语，将发现的结果作为句子的宾语从句，同时采用名词短语做同位语的语法结构，将全句的内容和结构进行紧凑衔接。

译文：The careful study of spectral lines shows that many of them actually consist of two or more close lines, something that the Bohr theory cannot account for.

◇ 例3

原文：功率因子为 60% 意味着，每消耗 600 W 功率就得提供 1000 VA 的功率，这种情况是我们想要避免的。

原译：A power factor of 60% means that we need to supply 1Kva of power for the consumption of every 600 watts of power, which is a situation that we want to avoid.

解析：原译添加了人称主语，将"提供"翻译成动词，处理成主谓宾结构的句型，同时后半个分句使用了非限定性定语从句来指代前句。原译表意正确，但结构较为松散，需要额外添加人称词。本句可采用逐词翻译，将"提供"处理成名词，并使用名词短语作同位语的结构进行翻译使句子结构紧凑。

译文：A power factor of 60% means that the consumption of every 600 watts of power needs the supply of 1Kva of power, a situation that we want to avoid.

◇ 例4

原文：软件定义网络(SDN)是一种新兴的网络架构，可以将网络控制平面与数据平面分离，从而实现网络资源的灵活管理和配置。

原译：Software Defined Networking (SDN) is an emerging network architecture. It can separate the network control plane from the data plane, and enable flexible management and configuration of network resources.

解析：原译将第一个分句翻译成独立的句子。然后将第二个分句和第三个分句用并列句的方式连接起来，这使三个分句被割裂开来，并且句子的结构单一。本句可采用逐词翻译方法，将第二个分句用定语从句的形式对"网络架构"进行修饰，并用分词状语将第三个分句的意义进行关联。

译文：Software Defined Networking (SDN) is an emerging network architecture that separates the network control plane from the data plane, enabling flexible management and configuration of network resources.

◇ 例 5

原文：物联网(IoT)是一种互联设备的网络，使得不同类型的设备可以相互通信和交互，并为用户提供更智能的服务和体验。

原译：The Internet of Things (IoT) is a network of interconnected devices. It allows different types of devices to communicate and interact with each other, and provides users with more intelligent services and experiences.

解析：原译将第一个分句独立出来，用并列结构将后面两个分句进行衔接，割裂了三个分句的联系。可以运用定语从句及动词分词的语法结构进行句型基本结构的逐词翻译，不需要再额外增加人称主语。

译文：The Internet of Things (IoT) is a network of interconnected devices that allows different types of devices to communicate and interact with each other, providing users with more intelligent services and experiences.

练习
practice

1. 简答题。
(1) 逐词翻译的定义是什么？
(2) 逐词翻译有什么优点和缺点？
2. 翻译下列句子，尝试采用逐词翻译法。
(1) 软件应用程序需要至少 4 GB 的 RAM 才能运行。
(2) 用户界面已经重新设计，具有新的外观和特点。
(3) 键盘有内置触摸板以便导航。
(4) 设备有内置摄像头，具有变焦和自动对焦功能。
(5) 该软件兼容 Mac 和 Windows 操作系统，该设计扩大了其应用领域。
(6) 这个人工智能系统可以自动学习和优化，提高其性能和准确度。
(7) 云计算平台可以提供强大的计算和存储能力，用于大规模数据处理。
(8) 这个机器人可以自主地行走和执行任务，不需要人类干预，这种情况以前只出现在我们的想象里。
(9) 数据库里存储了大量的信息，可以给研究者提供丰富的数据。
(10) 这个虚拟现实系统可以给用户沉浸式体验，让他们感受到仿佛身处不同的场景中。

第二节 直译翻译法

　　直译是指将一种语言中的词汇、语法、表达方式等直接翻译到另一种语言中，不进行任何调整和改变，以保留原有语言的特点和风格。直译强调按照译语规范，再现源语的词汇、句法结构和词的惯用法，而上一节所讲的逐词翻译是指将源语言文本的每个单词，按照一定的语法规则直接翻译成目标语言文本，虽忠实于原文，但对翻译结果的可读性有一定影响。直译和逐词翻译的区别在于：直译是在保持原文形式和结构的基础上进行翻译；逐词翻译则是紧跟原文进行翻译，可能会破坏目标语言的语法结构和习惯用法。直译并不一定等于逐字逐句翻译。逐词翻译虽然传达了原文的意思，但是常常语序奇怪，读起来不太自然。因此，在翻译过程中，需要根据翻译目的和语言习惯进行适当的调整和平衡，确保翻译的准确性和表达的流畅性。

　　在科技领域的翻译中，为了更准确地表达作者的意思并让读者易于理解，常常需要对逐词翻译进行调整和改变，以符合目标语言的语境和语法规则，这就是直译翻译法。直译翻译法相比逐词翻译法应用更普遍，也是科技英语翻译所采用的常规方法。

【经典赏析】

原文：机器学习是人工智能的一个重要分支，它可以帮助计算机系统自动从数据中学习，并改进其性能。机器学习的应用非常广泛，包括语音识别、计算机视觉、自然语言处理等。目前，机器学习已成为许多领域中的核心技术，持续不断地带来了新的突破和变革。

译文：Machine learning is an important branch of artificial intelligence that can help computer systems learn automatically from data and improve their performance. The applications of machine learning are very wide-ranging, including speech recognition, computer vision, natural language processing, etc. Currently, machine learning has become a core technology in many fields, continuously bringing new breakthroughs and changes.

赏析：译文采用了直译的翻译方法，高精准度地传达了原文的丰富内容，将具体应用范围一一列出，避免了意思丢失和歧义等现象，使翻译结果更加准确。在处理分句衔接时，没有受限于逐词翻译方法的限制，比如"它可以帮助计算机系统"该分句，在译文中用定语从句避免了不必要的主语重复。

【案例解析】

◇ 例 1

原文：目前，我国电子信息通信工程领域取得了巨大的进步，在 5G、人工智能、区块链等前沿技术上处于领先地位。

原译：Currently, our country's electronic information communication engineering field has made huge progress and in the aspects of cutting-edge technologies such as 5G, artificial intelligence, and blockchain, China is the first.

解析：原译更偏于逐词翻译，出现了两个问题：一个是"电子信息通信工程领域取得了巨大的进步"出现了 field 作主语的句子，表达生硬且不符合目标语的表达习惯；另一个是"在 5G、人工智能、区块链等前沿技术上处于领先地位"的翻译在前，而把"处于领先地位"的翻译放后，导致重点表意内容不够突出。

译文：Currently, China has made huge progress in the field of electronic information communication engineering and is in a leading position in cutting-edge technologies such as 5G, artificial intelligence, and blockchain.

◇ 例 2

原文："中国天眼"是国家重大科技基础设施，实现了我国在前沿科学领域的一项重大原创突破。

原译：The Five-hundred-meter Aperture Spherical Radio Telescope (FAST) is a major scientific and technological infrastructure facility, realizing major progress in innovation the country has made in cutting-edge technologies.

解析：原译前半句用逐词翻译方法译出，后半句采用动词分词短语，并在逐词翻译的基础上做了语序调整。原译的问题在于 realizing…部分的表达：首先，重要信息处理为分词短语不能凸显其重要性；其次，"一项"未被翻译出来，译文和原文信息不对等；此外，realizing 的逻辑主语无论是 The Five-hundred-meter Aperture Spherical Radio Telescope 还是其所在的整个句子，都不合理。因而本句在直译中，根据目标语的表达规范，可增译 as 作为介词引出"基础设施"，从而将原文后半部分译为句子主干，凸显重要信息，并将"一项"巧妙译为 an example。

译文：The Five-hundred-meter Aperture Spherical Radio Telescope (FAST), as a major scientific and technological infrastructure facility, is an example of the progress in innovation the country has made in cutting-edge technologies.

◇ 例3

原文：中国在信息产业领域拥有丰富的人才资源和优越的技术设备，已经成为国际上备受关注的电子信息大国。

原译：China has rich talent resources and superior technical equipment in the field of information industry, and has become an electronic information power in the world.

解析：原译采用直译翻译方法，前面分句根据目标语表达规范进行了语序调整，后半句将"国际上备受关注"简单处理成了 in the world。而本句中"国际上备受关注"可以用复合短语表达出来，直接作"电子信息大国"的修饰语，能更直观地体现直译翻译法的运用。

译文：China has rich talent resources and superior technical equipment in the field of information industry, and has become an internationally recognized electronic information power.

◇ 例4

原文：人工智能语音交互技术可以使机器与人类进行自然对话，实现智能化的人机交互。该技术的目标是实现高效、自然、智能的交互方式，提高人机交互的效率和质量。随着技术的不断进步，语音交互技术已广泛应用于智能音箱、智能手机、智能家居等场景中，为人们提供更加智能、便捷的生活体验。

原译：Artificial intelligence voice interaction technology can enable machines to have natural conversations with humans and it can achieve efficient, natural, intelligent human-machine interaction, and improve the efficiency and quality of human-machine interaction. With the continuous advancement of technology, voice interaction technology has been widely applied to scenarios such as smart speakers, smartphones, and smart homes, providing people with smarter and more convenient life experiences.

解析：译文将"实现智能化的人机交互"与"实现高效、自然、智能的交互方式"通过组合译的翻译技巧进行了内容和信息的简化，用一个动词短语包含进去了原文的两个"实现"动词。本句中，"该技术的目标是……"

是一句明确点明技术目标的核心句，适合采用直译的方式，利用不定式作表语的语法结构。

译文：Artificial intelligence voice interaction technology can enable machines to have natural conversations with humans and achieve intelligent human-machine interaction. The goal of this technology is to achieve efficient, natural, and intelligent interaction, and improve the efficiency and quality of human-machine interaction. With the continuous advancement of technology, voice interaction technology has been widely applied to scenarios such as smart speakers, smartphones, and smart homes, providing people with smarter and more convenient life experiences.

练习 practice

1. 简答题。

(1) 直译与逐词翻译有什么区别？

(2) 直译能达到什么样的翻译效果？

2. 采用直译的翻译方法翻译下列句子。

(1) 人工智能(AI)技术包括机器学习、深度学习和自然语言处理等领域，可以通过训练算法实现识别图像、语音和文字等任务。

(2) 电子信息通信工程领域的专业人才需要具备全面的知识和技能，将学科融合和跨领域应用作为发展方向和目标。

(3) 随着人工智能技术的不断提升和应用，电子信息通信工程领域将迎来更广阔的发展空间和应用前景。

(4) 电子信息技术的广泛应用和发展对工业生产、军事安全、公共服务等领域产生了深刻的影响。

(5) 信息安全风险等挑战和问题不断出现，需要通过技术和管理手段加以解决，确保通信和数据的安全可靠性。

(6) 计算机网络技术在电子信息通信工程领域的应用十分广泛，对于通信网络的互联和数据传输具有重要的意义。

(7) 计算机领域的教育培训是电子信息通信工程领域发展的重要保障和基础，需要依靠高水平的师资队伍和先进的教学设施。

(8) 云计算、大数据等先进技术的出现和应用，让物联网和共享虚拟空间领域迎来了前所未有的发展机遇和挑战。

(9) 全球定位系统(GPS)是一种卫星导航系统,该系统可以通过卫星发射的信号来确定地面上的位置和时间。GPS 技术已经被广泛应用于交通、航空、军事等领域,可以提高位置信息的精度和可靠性,为人们提供更加便捷和安全的出行体验。

(10) 云计算是一种基于互联网的计算方式,可以实现分布式计算和数据存储。通过 云计算技术,用户可以快速弹性地部署和管理计算资源,提高资源利用率和灵活性。目前,云计算已经成为许多企业和个人使用的重要技术,促进了数字化转型和信息化建设。

第三节　意译翻译法

意译就是只保持原文的内容,不保持原文的形式的翻译方法,而直译是最大程度既保持原文的内容,又保持原文的形式的翻译方法。具体而言,一方面,在内容层面上,译文要做到忠实于原文,这不仅是直译翻译方法的基本要求,也是任何翻译要达到的要求;另一方面,在形式上,直译翻译法要求译文要保持和原文一样的形式,不随意更改原文的形式。而意译不需要在形式上与原文追求一致,只需要忠实地将原文内容传达出来即可。也就是说,译文不用拘泥于原文的形式来传达原文的意思。

在实践中,直译和意译各有其应用的场合和效果。如果译文和原文用了相同的形式,并能表达出同样的内容,在此基础上,原文和译文还能产生同样的效果时,译者就应该选用直译的翻译方法。而意译是从整个句子所表达的意思出发来进行语言的对等翻译。这种翻译不是很注重形式层面的对等,而是注重整个翻译功能的实现。所以,意译所注重的是功能层面的对等。在科技文本翻译的过程当中,不仅需要不断发展字面意思的表达,还需要逐步推进相关功能的实现。不过,需要指出的是,由于意译存在着一定的主观性,在具体应用过程中并不一定优于直译,译者还是要根据具体情景来具体判断。

【经典赏析】

原文:5 月 30 日,搭载神舟十六号载人飞船的长征二号 F 遥十六运载火箭在
　　　酒泉卫星发射中心点火发射,3 名中国航天员顺利进入太空,将在空间
　　　站开始 5 个月的太空工作和生活。

译文:China launched the Shenzhou-16 manned spaceship on May 30, sending three
　　　astronauts to its space station combination for a five-month mission. The

spaceship, atop a Long March-2F carrier rocket, blasted off from the Jiuquan Satellite Launch Center in northwest China.

赏析：本句在翻译时采用意译法突破原句结构，第一句话以 China 为主语，凸显我国的重大事件，以 sending 将原文第二句话转化为状语，将最后一句话转化为 for 引导的目的状语。随后另起一句，将原文第一句话中的技术性信息作为补充说明介绍给读者。整句翻译细节处理到位，翻译灵活。

【案例解析】

◇ **例1**

原文：12 月 21 日，国务院新闻办公室发布《新时代的中国能源发展》白皮书。白皮书介绍了新时代中国能源发展成就，全面阐述了中国推进能源革命的主要政策和重大举措。

原译：On Dec. 21, the State Council Information Office released the white paper, titled "Energy in China's New Era", which introduces the country's achievements in energy development in the new era and comprehensively states major policies and measures for energy reform.

解析：原译采用了直译的翻译方法，将"国务院"作为句子主语，用过去分词的形式引出白皮书的题目，并用定语从句介绍了白皮书的主要内容。这样翻译使得本句想表达的核心主旨"白皮书"退居二线，没有体现出白皮书才是本句最重要的介绍点，应该把"白皮书"作为本句主语，凸显其重要地位。同时，"介绍"和"阐述"在英文表述中略显重复和冗长，可以用 a full picture of 将白皮书的两项内容介绍进去，简化表达。

译文：The white paper, titled "Energy in China's New Era", was released by the State Council Information Office on Dec. 21 to provide a full picture of the country's achievements in energy development in the new era and major policies and measures for energy reform.

◇ **例2**

原文：国家能源局日前表示，我国能源行业将聚焦能源安全保障、清洁低碳转型、科技自立自强、体制机制创新、加强国际合作五个方面，推进能源行业高质量发展。

原译：China's National Energy Administration currently said that the country's energy industry would focus on five aspects such as ensuring energy security, clean and low-carbon transformation, self-reliance in science and technology, system and institutional innovation, and enhancing international cooperation to advance the high-quality development of the energy sector.

解析：原译采用直译的翻译方法，句子过长，可读性较差。原译将"推进能源行业高质量发展"这一重要信息按照原文语序保留在最后，重点不突出。在细节方面，将"表示"译为 said that，语气不够强，语体不够正式，语言不够简洁；"能源安全保障、清洁低碳转型、科技自立自强、体制机制创新、加强国际合作"五个并列短语有的为-ing 结构，有的为名词短语，不符合并列短语写作规范；"体制机制"逐词翻译，不够简洁。

译文：China's National Energy Administration has decided to make multi-pronged efforts to advance the high-quality development of the domestic energy sector. Such efforts mainly revolve around five aspects, namely ensuring energy security, boosting clean and low-carbon transformation, enhancing self-reliance in science and technology, facilitating institutional innovation and enhancing international cooperation.

◇ 例 3

原文：无线网络通过一系列信号塔来互联，如果飞越地面网络的乘客全在使用手机，网络可能会超载。

原译：Wireless networks are connected by a series of towers, if passengers flying over these ground networks are all using their phones, the networks could become overloaded.

解析：原文用逗号连接三个小句，原译也遵循了原句的组成结构，语言不规范。汉语常将状语置于句首，英文则常置于句末，原译未作调整，导致重点不够突出。可调整第二个小句和第三个小句的位置，使主句在前，状语从句在后，更符合英文表达习惯，并将这部分和第一个小句之间用句号、分号、连词 and 等连接。考虑到添加 and 会使句子略显冗长，建议使用句号或分号直接连接。

译文：Wireless networks are connected by a series of towers; the networks could become overloaded if passengers flying over these ground networks are all using their phones.

◇ 例 4

原文：人们早就怀疑，大气层中有一个"高温带"，其中心在距地面约三十英里的高空。利用火箭进行研究后，这一奇异的事实已得到证实。

原译：It had already been suspected that there is a "high temperature belt" in the atmosphere, with its center roughly thirty miles above the ground. After the rocket research, this strange fact has been confirmed.

解析：原译采用直译的翻译方法，保留了原句的两个独立分句进行翻译，句子结构和信息较为松散。可采用意译的翻译方法，通过合译的翻译技巧，利用 a fact 引导名词同位语的语法功能将两句合并，把"利用火箭进行研究"处理成句子的主语。

译文：Rocket research has confirmed that there is a "high temperature belt" in the atmosphere, with its center roughly thirty miles above the ground, a strange fact that had already been suspected.

◇ 例 5

原文：为了更好地感悟物理学课程，为今后深入学习打下坚实的基础，当你在学习中遇到物理学中的符号、公式、定义和定律时，不管它多么复杂都应该牢记心中。

原译：In order to understand the subject better and have a solid foundation for further study, when you come in contact with the symbols and formulae, definitions and laws of physics, no matter how complex they may be, you must fix in mind.

解析：原文中有多层限定条件，先陈述目的，再描述时间和前提，最后说明要做的事情。而英语的表达习惯完全相反，若按原层次翻译就会令人费解。本句可以采用意译的翻译方法以复合句译出，打破原句的句序和层次，主句先叙述应该干的事情，再从几个方面来补充说明主句。这样的层次排列符合英语的表达规范。正确的译法是将 fix in mind 先译出，然后译出 when 从句和 no matter how 从句，最后用 in order that 给出目的。

译文：You must fix in mind the symbols and formulae, definitions and laws of physics when you come in contact with them, no matter how complex they may be, in order that you may understand the subject better and have a solid foundation for further study.

在实际的科技翻译中，并不会将翻译方法割裂开来，每句话的翻译只能采用一种翻译方法，也并不是只用到本章所介绍的三种翻译方法，而是将八种翻译方法综合运用，其中以本章所介绍的三种翻译方法最为常见。综合翻译就是根据科技文本的不同来进行具体的实施，翻译时要求译者相对比较灵活地运用翻译方法。在科技笔译过程中，翻译方法有较大的提升与发展空间。总之，科技文本的翻译方法是多元的，在科技文本翻译的过程中，需要以科学、灵活的方法来具体展开翻译工作。只有这样，科技笔译的翻译水平与翻译效果才会得到质的提升。

练 习
practice

1. 简答题。
(1) 意译和直译的区别是什么？
(2) 使用意译翻译法应该注意什么？
2. 赏析下列翻译，指出译文的优点。
(1) 近日，中共中央、国务院印发了《数字中国建设整体布局规划》(以下简称《规划》)。《规划》指出，建设数字中国是数字时代推进中国式现代化的重要引擎，是构筑国家竞争新优势的有力支撑。

China has rolled out a plan for the overall layout of the country's digital development. Building a digital China is important for the advancement of Chinese modernization in the digital era, and provides solid support for the development of new advantages in the country's competitiveness, according to the plan, which was jointly released by the Communist Party of China Central Committee and the State Council.

(2) 晶体管有两大优点：一是体积小，二是能密集排布而不致过热。

Two of the advantages of the transistor are its being very small in size and its being able to be put close to each other without being overheated.

(3) 科技创新的飞速发展，催生出一批又一批创业公司，借助技术手段和商业模式，不断探索新的商业领域和经济模式，为经济增长注入新的动力和活力。

The rapid development of technological innovation has spawned a series of entrepreneurial companies that continuously explore new business fields and economic models through technical means and business models, injecting new impetus and vitality into economic growth.

(4) 在未来的 3～5 年间，公司在全国各地将会拥有超过 30 个转运仓库，并在地方组建本地运输团队，从而真正建成能够控制终端配送质量的全国性物流网络。

In the next three to five years, we all have more than 30 transit warehouses throughout China and set up a local transport team, providing an even greater national logistics network with the same high quality.

(5)　"蓝芯"重磅车型荣威 360 在空间、油耗、动力、配置等多方面均拥有足以抗衡合资竞品的实力。

The ROEWE 360, a star product of the Net Blue series, is able to contend with similar joint venture cars in terms of space, fuel consumption, engine and configuration.

3. 使用意译翻译法翻译下列句子。

(1)　最近，使用专门设计的电子显微镜已经能拍摄重原子照片，只是非常模糊。

(2)　由于其快速、高效、低成本的优势，3D 打印技术在医疗、工业制造等领域得到广泛应用，在医疗领域中越来越多地被用于建造各类人体器官模型。

(3)　通过无线传输技术的应用，实现设备之间的远程通信和数据交互，在智能家居、物联网等领域中发挥着重要作用。

(4)　在区块链技术的应用场景中，因其本身不可篡改的属性以及去中心化的优势，使其在金融业领域中的应用尤为广泛。

(5)　通过多次迭代和优化设计，该算法在兼顾效率和可靠性的基础上，较为有效地减轻了人工干预的需求，并在一定程度上提高了机器匹配的准确率。

第七章　科技翻译工具

工欲善其事，必先利其器。许多人认为进行翻译所需要的一切就是 Microsoft Word，几本词典和译者所掌握的语言知识。事实上，为了提高翻译的效率和质量，专业译者在翻译过程中需要利用多种翻译工具。广义的翻译工具包含翻译各个环节所涉及的各种硬件和软件，其中硬件包括纸、笔、词典、计算机、扫描仪、手机等；软件则涉及光学字符识别(OCR)软件、字处理软件、文档阅读与编辑软件、文档格式转换工具、搜索引擎、语料库、计算机辅助翻译(CAT)工具等。狭义上的翻译工具主要指计算机软件工具。

近年来，人工智能技术的突破推动了翻译技术和工具的迅猛发展，翻译行业出现了诸多新的变化。翻译的模式大致会经历人工翻译、人译机助、机译人助以及智能机译等几个阶段。当前正处于由人译机助逐渐转向机译人助的过渡过程。从传统笔译到机器翻译加译后编辑模式，从传统的口译到机器辅助口译，人机交互程度越来越高。翻译工具也逐渐从单一功能升级到多功能或全功能，覆盖用户多元化、多层次的翻译需求。翻译技术提供商将词典查询、翻译搜索、记忆匹配、机器翻译、译后编辑、项目管理、质量评估等功能无缝整合，提供了一体化的翻译服务体验。未来，高度智能化的翻译系统将会自动连接翻译所需的一切资源，充分发挥人工智能赋能的优势，将译者的智慧聚焦在更有价值和创造性的工作中，在大幅度降低翻译的劳动工作量的同时提高翻译效率。

过去，译者凭借自身硬功夫以单打独斗的方式获得翻译市场的竞争优势，但这种服务方式逐渐被人机结合的服务模式取代。如今，为了具有真正的竞争力，无论是翻译机构的译者还是自由译者都有必要了解并掌握翻译软件工具，尤其是 CAT 工具的使用方法和技巧。

第一节　翻译工具概述

翻译过程中涉及多种工具的应用。根据用途，翻译工具可以粗略地分为两大类：通用工具和翻译专用工具。

1. 通用工具

通用工具，顾名思义，除了翻译行业也应用于其他领域，比如常见的办公软件 WPS、Microsoft Office、PDF 阅读器和编辑器 Acrobat、文档格式转换工具等即属于此类。其中和科技翻译关系较为紧密，值得科技译者重点掌握的主要有以下三类。

1) 词典工具

相较于厚重、翻查不便的纸质词典，电子词典软件可以同时加载并查询多款词典的词库(如权威的《牛津高阶英汉双解词典》《朗文当代英语词典》《柯林斯高级英语词典》《韦氏词典》等)，不仅便于译者对比参照，而且可以节省查询时间，提高查词效率。一些 CAT 平台，如商用的 SDL Trados 和开源的 OmegaT 直接集成了词典功能，支持自行加载用户词库，译者可以直接在翻译编辑界面进行取词查询、复制粘贴释义，不必另行打开词典软件，进一步提升了效率。目前流行的免费离线词典软件有灵格斯词典、有道词典、欧路词典等，在线词典平台包括有道、必应、Linguee 等。

但就汉英科技翻译而言，词典工具的应用场景较为有限。一方面，离线的字典词库和纸质词典通常具有信息滞后局限性，即收录的科技词汇有限，很多时候查不到新近出现的术语，抑或已有的词汇并未收录其最新用法和释义，跟不上日新月异的科技进展。另一方面，汉英科技词典通常只提供对应的英文名词术语(有些词典也提供语域信息，比如标注其是计算机、机械等领域的术语)，而不提供汉英平行对照的例句，译者无法保证其在特定上下文中用法的准确性和地道性。相比之下，在线词典的词库通常更新比较及时，也有汉英对照例句可供译者参考，在一定程度上可以弥补这两方面缺陷。然而鉴于目前权威汉英科技平行语料库的缺乏，这些例句有些直接来自互联网上良莠不齐的语料，有些甚至是未经人工审查的机器译文，因此同样无法保证这些例句的权威性和地道性。科技译者仍需查询相关的权威科技资料，如科技专著、教材，尤其是最新的相关论文，以寻求准确可靠的参考例句。

2) 语料库工具

语料库指经科学取样和加工的大规模电子文本库，借助计算机分析工具，研究者可以开展相关的语言理论及应用研究。语料库是语料库语言学研究的基础资源，也是经验主义语言研究方法的主要资源，主要应用于词典编纂、语言教学、传统语言研究、自然语言处理中基于统计或实例的研究等方面。高级翻译人员也常常使用语料库来了解词语、句式在母语国家的使用率、使用背景和分布，使自己的翻译更加地道纯正。平行语料库(parallel corpus)也称双语语料库,对于双语词典编纂、机器翻译和 CAT 有重要意义;可比语料库(comparable corpus)和单语语料库在科技翻译实践中扮演着不可或缺的角色。语料库在提供鲜活、真实、地道的语言示例方面格外有用,字典和搜索引擎提供的细节和上下文远不及语料库。语料库的构建、检索和在翻译中的应用将在下一节进行详细讲解。

3) 资料查找工具

在科技日新月异的今天,再聪明的大脑也难以存储海量的专业知识。因此,译者必须具备良好的信息检索、辨析、整合和重构能力。在查找资料时明确要找什么,去哪里找,用什么工具找,这是信息化时代译者应该具备的基本能力。如何在有限时间内从浩如烟海的互联网中找到所需信息,如何通过专业语料库验证译文的准确性等,皆需要借助信息检索能力。因此,当代译者应熟练掌握主流搜索引擎和语料库检索工具的特点、搜索关键词的选择、检索语法(如通配符和正则表达式)的使用、权威语料资料来源等,以提升检索的速度和检索结果的质量。

2. 翻译专用工具

与通用工具相对的是专门为翻译工作开发的工具,主要指各类计算机辅助翻译(CAT)工具,它主要面向专业译员、兼职翻译人员和翻译专业的学生。从翻译各环节涉及的翻译工具核心功能角度,可以把 CAT 工具大致分为以下几类。

1) 翻译记忆(Translation Memory,TM)

翻译记忆是所有 CAT 工具的核心功能,也是译者使用最多的翻译工具。翻译记忆库中存储了已经完成的翻译项目源文本及其译文,可用于未来的翻译项目。翻译记忆由翻译单元组成,翻译单元通常是双语对照的短语、句子、段落或标题。翻译单元数量越多、质量越高,翻译记忆的功效越佳。

译者在 CAT 翻译系统内进行翻译时,翻译记忆工具在后台工作,把完成翻译的内容储存在一个双语库中,译者可以在以后完整地使用这些内容或者根据需要进行调整。它首先将待翻译的文本划分为称为"片段"(segments)的单元,随着翻译的推进,软件将已经翻译过的片段存储到数据库中。当软件识别到待译片段与记忆库中已译的片段完全匹配或高度匹配(可以设置匹配度阈值,如 75%)时,就会建议译者复用或参考已译片段的译文,从而避免重复劳动。

不同的 CAT 工具采用的匹配机制可能有所差异,具体取决于工具生产商采用的特定算法。一般来说,匹配有以下四种类型:

(1) 语境匹配(也称完美匹配):待译片段的语境和意思都与之前的翻译

100%匹配,即以往的项目中已经翻译过这个片段,CAT 工具也能识别其上下文片段。有些 TM 软件会将语境匹配标识为 101%或 102%匹配。101％匹配是指翻译记忆在与当前片段 100%匹配基础上,还与其前一片段或后一片段匹配;102％匹配则指不仅当前片段与记忆库存储完全匹配,而且其前一片段和后一片段也完全匹配。语境匹配的通常是标题、副标题和表格等。对于这种匹配,译者可以直接复用之前的译文,不需要进行任何改动。

(2) 100%匹配:待译片段与记忆库中的句子具有相同的词汇和格式,但上下文可能不同。对于这种匹配的句子,译员有必要留意检查一下,以避免不同上下文造成的差异可能被忽略的情况。但大多数情况下,译员不必对之前的译文进行调整或再次翻译。

(3) 模糊匹配:待译片段中的若干词汇与此前的翻译相匹配。这时可以用以前的译文来翻译匹配的部分,但必须翻译不匹配部分的内容。因为待翻译的内容可能跟记忆库有重合部分,因此不少翻译公司会按照匹配程度折扣计价。模糊匹配大多分为四个级别:95%～99%、85%～94%、75%～84%和 74%以下。前三档会有相应的匹配折扣,74%及以下普遍按无匹配翻译费率计算。例如,译员报价 200 元/千单词,若匹配度在 74%及以下,就按 200 元/千单词算;若匹配度为 95%～99%付费就会打个折扣,如 30%,那么译员实际得到的报酬就是 60 元/千单词。

(4) 无匹配:待译片段是全新的,必须从头开始翻译。

与文学翻译不同,科技文本的词汇和句式相对固定,译员积累一定规模的记忆库后,遇到高匹配程度的句子会越来越多,翻译工作将变得越来越轻松。

2) 术语数据库(Term Database,TD)

术语数据库是一个集成在 CAT 工具中的可搜索的数据库或词汇表。它可以存储单词或短语形式的术语以供检索使用,其功能类似于字典。与翻译记忆存储所有类型的单词不同,术语数据库专注于行业术语。可以把术语数据库当成是一种特殊的翻译记忆库,它有时也包含术语使用的参考说明和规则。翻译包含大量特定领域术语的科技文本时,译员可以从整合了术语数据库 CAT 工具界面直接得到术语的准确译文,不必费神在海量信息的互联网或字典中去搜索和斟酌。这样翻译的时间就可以大大减少,从而极大提高译员的工作效率。

3) 语言搜索引擎(Language Search Engine,LSE)

语言搜索引擎的功能和常见的互联网搜索引擎一样,但主要针对术语库、翻译记忆库、平行语料等进行精准高效的搜索,从而节省时间成本。同时,翻译记忆库的搜索结果还支持显示两个文本片段的精确匹配程度。当待译文本中的术语或句子在记忆库中没有匹配的译文时，可采用相关搜索功能从网络记忆库中寻找平行文本，通过比对平行文本可以得出恰当的译文。

4) 术语管理(Term Management)

与语言搜索引擎类似，术语管理工具使译者能够自动通过术语数据库(termbase)搜索，找到与待译文档中的术语相匹配的术语。在科技翻译中，几乎每篇文档都有大量的专业术语，译文术语前后一致是翻译校对的重要内容之一。这项工作费时费力，而且难免有疏漏，比如某个标书翻译项目有多名译员参与，对于"地铁"这个名词术语，有的译员译为subway，有的译为metro，还有的译为tube；若项目时间跨度大，同一个译员对"地铁"的译文也可能出现前后不一致。有了术语管理工具，就可以提前对术语进行规范。用户只需提前创建标准术语表(表中包括术语原文和译文)，在使用整合了术语管理工具的翻译记忆系统翻译时，系统会自动识别出当前句子中有哪些词是已定义的术语，并给出标准的术语译文。这样，术语的译文既精准又在整个翻译项目中保持前后一致，省掉了译后再次校对的麻烦。不同的翻译记忆工具对术语管理的具体实现方式虽然有差异，但最终都是为了保证术语的统一。

术语库中的术语可以提前利用术语提取工具从项目文本中提取并导入数据库，也可以在翻译的过程中由译员随时添加。翻译中常用的术语库一般为双语或多语种对照词汇表，有的术语库还支持为术语添加学科、行业、词性等标签以方便译员选择。

5) 交互式机器翻译(Interactive Machine Translation，IMT)

交互式机器翻译旨在通过译员与机器翻译引擎之间的交互作用实现人类译员的准确性与机器翻译引擎的高效性的有效整合。交互式机器翻译根据译员录入的翻译，动态提供后续的译文片段智能推荐和整句补全，供用户选取、调整或修改，然后根据译员作出的选择生成比初始候选译文更优的译文，再次供译员反馈，如此循环，直至生成满意的译文。交互式机器翻译能够在翻译过程中节省大量的手动词典查询和文字输入工作量。结合翻译记忆工具，交互式机器翻译还能够记住译员对错误的机器翻译结果进行的修正，从而避免传统机器译文译后编辑中重复多次进行人工修订。

6) 翻译对齐(Translation Alignment)

对齐就是将整理好的原文与译文以句子为单位进行对齐排列，生成翻译记忆双语(或多语)文件，并以翻译单元(源语言和目标语言句对)的形式存储，可以直接用于翻译记忆库，供日后使用。对齐的双语(或多语)对照文本也有利于后续的翻译复核以及质量审查，避免阅读和比较两个甚至多个单独文件的麻烦，可为审校人员节省大量的时间和精力。翻译对齐工具是创建双语平行语料库(即翻译记忆库)必不可少的工具，而翻译记忆库的体量和质量直接影响着 CAT 软件的使用效能。CAT 新手可以利用对齐工具将自身之前的翻译成果转化成翻译记忆文件，也可以获取网上公开的平行语料库直接转化为翻译记忆库。

练习 practice

1. CAT 有哪些核心功能？
2. 翻译记忆的基本原理是什么？

第二节　语料库工具

在科技翻译过程中要进行大量的资料查阅，语料库是必不可少的工具。语料库是语料构成的集合，语料库的构建、处理和应用离不开相应的软件工具。常用的语料库工具有以下几类。

1. 语料采集工具

构建语料库所需的语料来源多样，如网页、电子文档、纸质材料等，在采集不同来源的语料时需要使用不同的工具。OCR 识别工具可用来处理纸质文档或图片版 PDF 等不可编辑的文档，将之转换为可编辑的文本文档或 Word 文档；网络爬虫类工具，如八爪鱼、Web Scraper、You-Get 等，可以自动批量采集网页上的语料资源。对于 Epub、Mobi 等电子格式的书籍，可以用 Calibre 等软件或在线格式转换工具将之转换为 TXT 或 Word 格式来提取其中的文字信息。对于科技翻译而言，最常见的语料有相关领域的纸质或电子版专著、教材、论文等参考文献，电子版通常是 PDF 格式。

2. 语料清洗工具

语料清洗指去除语料中的噪音，包括不符合规范的文字符号、空格段落、标点符号等。通常利用 WPS 和 Microsoft Word 等支持通配符的文字处理软件进行查找、替换操作，即可进行普通的语料清洗工作。对于大批量、大体量的语料，可以使用支持正则表达式的文本编辑器进行批处理或正则替换，如 EmEditor、notepad++ 等。对于 html 等网页材料，可以用 TextForever 等软件进行处理。科技语料大多是标准的 pdf 文档，如果不进行深层次的语料库语言学研究，只是从中获得语言表达参考，则通常不需要进行清洗。

3. 语料对齐工具

语料对齐一般指的是双语或多语文本的平行对齐，目前应用最为广泛的语料对齐是句级语料对齐。常见的语料对齐工具分为两大类：一类是 CAT 平台自带的工具，如 Trados WinAlign 和 MemoQ LiveDocs；另一类是独立的工具，如 ABBYY Aligner、LF Aligner、TMX Editor、Tmxmall 在线对齐等。一般来说，CAT 自带的语料工具功能有局限，独立的专用语料对齐工具功能更强大、效率更高、效果更好。

4. 语料检索工具

语料检索指的是对语料中的词、句或结构进行检索，可以获得词频、词密度、词表、搭配、关键词单等，以便对语料进行分析研究。大型在线通用语料库一般都提供了配套的检索工具。此外，还有一些专用的语料检索和分析软件，如 AntConc、ParaConc、Corpus Workbench(CWB)等，可以用于对自建的本地语料库进行检索分析，但这些专用软件通常要求语料必须是清洗过的 UTF-8 编码 TXT 格式文本文档。对于科技翻译而言，通常无须对语料进行高级的语言学分析与研究，自建小型语料库的语料多是排版精良的 PDF 格式科技专著、论文和资料，可以使用支持 PDF 文件检索的全文检索软件进行语料检索，如 Filelocator Pro、Everything 等，这样就可以直接使用 PDF 文件的语料而不用任何预处理。

翻译实践中常用的语料库类型有平行语料库和可比语料库。平行语料库是由原文文本及其平行对应的译语文本构成的双语/多语语料库，其对齐程度有词级、句级、段级和篇级几种，多用于双语词典编纂和翻译研究，双语平行语料库还可以作为翻译记忆直接用于计算机辅助翻译。可比语料库是由不同语言的文本或同一种语言不同变体的文本所构成的两个或两个以上的语料库。可比语料库可再细分为单语可比语料库和双语/多语可比语料库。前者收集同一种语言类似环境下的类似内容的文本，而后者收集的是在内容、语域、交际环境等方面相近的不同语言文本，多用于对比语言学。单语和双语可比语料库在科技汉英翻译实践中都有广泛的应用。

双语语料库对翻译的重要性体现在：便于从语言对比中深入了解所对比的语言，而这种了解往往容易在研究单语种语料库时被忽略；通过一系列比较，揭示语言的共性以及某语种所特有的、语言类型与文化上的差异，以及原文与译文、母语与非母语之间的差异。

就翻译实践而言，语料库为译者提供了一个工作平台和参考工具，可用来提高译者的语言及文化意识。在这方面，即使是单语语料库也很有用。Bowker(2001)发现，在对专业领域的理解、术语的准确选用及习惯表达等方面，用语料库辅助的译文比用传统资源(如只靠词典)的译文质量更高。使用单语语料库，译者可以从更多的句子、语境中理解待翻译原文，并利用语料库统计分析结果中选择合适的译文。汉英科技翻译实践中，实用性最强的也是单语语料库，这是因为符合翻译项目的特定领域双语语料库几乎找不到。英语单语语料库可及性最佳，不仅有很多成熟的大型语料库供译者免费获取，也很容易从互联网上获得大量高质量免费语料来自建单语语料库，同时也有成熟的检索工具可供使用。英语语料库网(www.english-corpora.org)整合了主要的流行英语语料库，如英国国家语料库(BNC)、美国当代英语语料库(COCA)、维基百科语料库、iWeb 语料库等。该网站提供了免费的在线检索界面，用户也可以选择购买和下载语料库至本地。权威的平行语料库有联合国平行语料库，由已进入公有领域的联合国六种工作语言的正式记录和其他会议文件组成，当前的 1.0 版本包含 1990—2014 年编写并经人工翻译的文字内容，包括以语句为单位对齐的文本，有现成的汉英双语文本和平行语料子库供下载。开放平行语料库网站 OPUS 提供了大量开源的平行语料子库，但其中汉英子语料库十分稀少，也没有专门的科技双语语料库，且语料的预处理都是自动进行的，语料对齐也未进行人工复查。

与单语种语料库相比，可比语料库对翻译研究的价值更大。研究表明，把外语翻译成母语时，参照可比语料库核对能够帮助译者提高表达能力，而且译文中的错误也会减少。把母语译成外语时，对语料库工具的需要成倍增加，而且语料库的用途也远远不止是核对母语中的理解难点，还包括专业背景知识获取、术语的准确选用及目标语习惯表达等。一个小型的可比语料库就可以有效提高译者理解原文，并用更流畅的目标语进行翻译的能力。在科技汉英翻译实践中，因为译者通常对待译的科技文本较为生疏，所以根据翻译任务涉及的特定专业领域，利用上述语料采集工具搜集相关的中英文电子专著、教材、论文等语料，自建小型可比语料库就特别有必要。

需要指出的是，单语种语料库和可比语料库虽然都对翻译有用，却难以用这些语料形成对可能性译法的假设(Aston，1999)。而且，通过检索这类语料库来验证假设既费时又容易出错。与之相比，使用平行语料库更容易找到原文中特定表达方式的译法。因此，平行语料库可以帮助译者将术语和短语翻译得更加准确地道，从而得到了广泛的推荐。但在科技翻译实践中，很难获得可靠的科技汉英平行语料库。一方面原因是科技发展迅猛，技术信息迅速迭代，除非语料库是持续更新的，否则几年前的科技平行语料库可能已经过时，很难从中找到高匹配的文本；另一方面的原因是科技翻译涉及的范围十分广阔，细分领域更是数不胜数，通用型语料库或大型的学术英语语料库统计出来的分析结果并不一定适用于翻译项目所属的特定领域。因此，在进行科技汉英翻译时，目前只能退而求其次，以自建双语可比语料库和单语语料库为主，但可以利用有道、Linguee 等提供双语对照例句的在线词典作为可能译法的提示，然后在自建的可比语料库和单语语料库中对此译法进行验证，看真实地道的语料中是否确实有这种表达。如果没有检索到匹配的真实语句，则需对此译法进行改进。

练习
practice

1. 简答题。

(1) 语料库在翻译中有哪些主要用途？

(2) 常见的科技语料来源有哪些？

2. 在互联网上下载某电子产品的中英文说明书(通常为 PDF 文档)，利用恰当的工具提取其中的双语文字进行语料降噪，然后使用开源对齐软件 LF Aligner 进行语料对齐，生成一个翻译记忆 TXM 文档。

LF Aligner 4.25 下载地址：https://sourceforge.net/projects/aligner/。

3. 将自己以前翻译过的中英文对照文本进行语料对齐，制作成翻译记忆 TXM 文档。

第三节　计算机辅助翻译工具

进入 21 世纪以来，语言服务和本地化需求空前增长，多样化和专业化趋势日益凸显，待译内容和格式越来越复杂。CAT 技术使得繁重的手工翻译流程趋于自动化，大幅提高了翻译工作的效率和质量。随着语言技术研发的不断创新，翻译记忆工具(如 SDL Trados、MemoQ 等)、术语管理工具(如 SDL MultiTerm、crossTerm、TermStar 等)以及自动质量保证工具(如 ApSIC Xbench、Html QA、QA Distiller、ErrorSpy 等)等专业 CAT 工具相继出现，且功能愈加强大，逐渐成为现代翻译工作中应用最多的主流工具。

当前的 CAT 工具已经从单一功能升级到多功能或全功能，覆盖客户方多元化、多层次的翻译需求。翻译技术提供商将格式转换、翻译搜索、记忆匹配、内容推荐、机器翻译、译后编辑、项目管理等功能无缝整合，提供一体化的服务输出，典型代表是 SDL Trados Studio 2022。在一体化的 CAT 环境中，翻译工具和翻译环境高度融合，提供了多样化的工作方式，用户可以在线、离线或在集成解决方案中使用组合来工作。基于云的解决方案还提供了极大的灵活性，比如用户可以随时随地在网络浏览器中进行翻译或在平板电脑上进行审阅，而不仅限于桌面电脑工作场景。

CAT 软件也为各类型的机翻译引擎提供了广阔的应用场景，目前最常见的翻译模式就是 MT＋TM＋PE(机器翻译＋翻译记忆＋译后编辑)，SDL Trados Studio、memoQ、OmegaT 等主流 CAT 工具均提供了多种机器翻译引擎的 API 接口或插件(如 Google Translate、DeepL Translator、Bing Microsoft Translator 等)，以及质量评估、译后编辑、译前译后编辑差异等功能。越来越多的翻译技术提供商还开始使用翻译质量自动评估技术来辅助客户或语言服务提供商进行机器翻译引擎评估和选择、待译内容筛选、项目质量监控、译员筛选和培训等，如 ModelFront、语智云帆试译宝、一者科技的 YiCAT 等。

CAT 的主要优势有以下几点：

(1) 保持翻译一致性，包括相同句段的译文一致性、原文与译文格式一致性、术语翻译一致性。

(2) 提高翻译速度。CAT 一体化平台避免了重复翻译，语言资源的查找与管理方便，从而节省了大量烦琐的工作，提高了翻译效率。

(3) 方便语言资产的建立和维护。CAT 规范的翻译流程便于标准化生成和管理语言资产，语言资产库随着使用 CAT 工具的积累会变得越来越大和完善。这些资产可以用于多个项目，为日后的翻译任务提供更多的参考和支持，从而提高整体生产力和翻译水平。

(4) 降低整体翻译成本。虽然 CAT 软件的购买需要投入很多资金成本，熟练掌握 CAT 工具的使用也需要投入学习时间成本，但对于有大量复杂翻译任务的专业译员来说，翻译效率大幅提升带来的整体效益将远超初期的投入。

另外，CAT 工具支持几乎所有常见格式的文件，从 Word 到 HTML、XML、Open Office、PDF、Java Resource(Properties)、InDesign(INX/IDML)等。如果不使用任何辅助工具，译员只能在现成的可编辑文件上工作，如 Microsoft Word 或 Microsoft Excel。鉴于越来越多的客户要求翻译更复杂的文件类型，不会使用 CAT 工具的译员可能会丧失承接这类翻译任务的机会。

当然，CAT 也有其局限性，它的主要缺陷是：基于句段的翻译模式割裂了原文的连续性，造成翻译中调整句序困难；逐句段翻译习惯可能导致译员缺乏上下文观念，对机器译文和匹配译文过度依赖还会产生惰性；模糊匹配译文的可用性难以保证；应用有局限，在处理复杂句子、俚语和口语方面仍然存在限制和误差，尤其是高度技术性或文化方面的内容仍然需要人类译员进行编辑和修正，对于文学翻译基本无能为力；CAT 软件成本和学习时间成本较高。

对于使用翻译工具的新手，建议如下：

(1) 选择适合自己的 CAT 软件。市面上有很多不同的 CAT 软件，如商用的 SDL Trados、memoQ、开源的 OmegaT 等。在选择软件之前，需要先了解自己的需求，比如要翻译什么类型的文档、需要与哪种格式的文件兼容、考虑翻译项目规模大小、CAT 软件投入产出比等。兼职的自由译者推荐开源免费的 OmegaT；翻译工作量巨大的专业译员最好还是选择业界通用的商用 CAT 工具或平台；翻译专业的学生则应尽量多尝试几种流行的 CAT 工具，对于商用 CAT 软件可以利用其免费试用版(通常为期限为一个月)入手。

(2) 学习软件操作基础。CAT 软件需要学习和练习才能掌握。新手可以通过官方教程、在线视频等方式学习软件的基础操作，例如如何导入文档、创建翻译记忆库等。

(3) 建立自己的翻译记忆库。CAT 软件最核心的功能就是翻译记忆库。建议将自己翻译过的文档进行语料对齐做成翻译记忆，导入 CAT 记忆库中，以便在以后翻译类似文档时可以重复使用之前翻译过的内容，提高翻译效率并减少错误率。

(4) 充分利用机器翻译。CAT 软件中的整合机器翻译可以提高翻译效率，但要注意机器翻译的结果并不总是准确的。建议在练习时尝试使用机器翻译，但在最终翻译产品时仍然需要进行人工编辑和校对。CAT 软件整合的机器翻译服务通常会按调用接口翻译的字符数收费，用户需提前注册并购买相应的机器翻译接口服务，并在 CAT 软件中进行配置。一般会有一定的免费额度供用户试用，如腾讯机器翻译，其文本翻译的每月免费额度为 500 万字符，语音翻译的每月免费额度为 1 万次调用，但文件翻译则无免费额度。对于字数不多的文档，也可以利用网页版机器翻译，大部分机器翻译引擎都提供 5000 字符以内的免费翻译服务。

(5) 不断实践和积累经验。只有不断地使用和练习 CAT 软件才能掌握它的各种功能和技巧，提高翻译效率和质量。建议多参与翻译社区或者小组，分享经验和学习心得，不断提高自己的翻译水平。

练习
practice

1. 简答题。

(1) 主流的 CAT 工具通常包含哪些功能模块？

(2) CAT 的优点和缺点分别有哪些？

2. 利用商用 CAT 工具 Trados Studio 2022 翻译一个科技汉语内容的 HTML 网页。(访问 https://www.trados.com/，下载安装 Trados Studio 2022 免费试用版。仔细阅读产品简介，观看官方视频教程 (https://www.bilibili.com/video/BV1ks411W7rZ/)，完成翻译项目创建和机器翻译引擎配置。)

3. 利用开源 CAT 工具 OmegaT 翻译一篇电子产品的汉语说明书。(访问 https://omegat.org/zh_CN/download，下载安装 OmegaT 6.0.0，仔细阅读用户手册 PDF 文档。完成翻译项目创建，导入上一节创建的产品说明书 TXM 文件作为翻译记忆库。)

第四节　机器翻译引擎

机器翻译(machine translation)也称为自动翻译，属于计算语言学的范畴，是利用计算机程序将一种自然语言(源语言)自动转换为另一种自然语言(目标语言)的过程。随着经济全球化与互联网的飞速发展、计算机算力的提升和多语言信息的爆发式增长，机器翻译技术日渐成熟，尤其在近些年取得了长足进展，机器翻译的实用价值日益凸显，在促进政治、经济、文化交流等方面起到越来越重要的作用。

机器翻译经历了基于词典替换方法、规则的方法、实例的方法、统计的方法和神经网络系统技术的发展阶段。目前最常用的机器翻译是基于统计和神经网络的技术，其运作原理相似，都是用基于一组统计数学或神经网络算法进行编程，形成机器自主学习的模板或逻辑，然后经过海量语料的学习，提炼双语对应的映射规则。每一次翻译都是机器在无数条翻译规则中寻找一条最为适配的规则，是对语料的拆分、重组而产出的结果。

随着 AI 技术的提高，神经网络机器翻译(NMT)发展迅速。神经网络机器翻译使用人工神经网络，通过使用单个整合模型对完整句子进行建模，从而预测字词序列翻译后的可能性。当前应用最广的机器翻译引擎，如 Google

Translate、DeepL Translator、有道翻译、百度翻译等，都是基于神经网络的机器翻译技术。目前 NMT 除了可以翻译文本外，还具备语音翻译、实物翻译、涂抹翻译、离线翻译等功能。NMT 将强大的机器翻译技术和语音识别、语音合成、图像识别、文字识别等完美结合，创造了绝佳的用户体验，比如一个完全不懂外语的游客，完全有可能在手机拍照翻译和涂抹翻译的帮助下顺利地在异国点餐和获得交通指引。在市场需求的推动下，商用机器翻译系统迈入实用化阶段。基于目前的人工智能技术发展，虽然机器翻译在很长一段时间内还无法完全取代人工翻译，但基于海量学习数据的先进机器翻译技术已经突破了传统基于规则和实例的机器翻译局限，并在多个场景下成功解决了用户的具体需求。对那些输出结果的质量不是很重要的文本翻译而言，机器翻译已经开始取代人工翻译完成简单翻译任务。例如，如果用户只是想了解一份技术文档的一般内容和信息，而不关心译文是否译得漂亮，那么机器翻译生成的可理解的译文就足够满足用户需求了。

机器翻译结果的好坏，往往取决于译入及译出语之间的词汇、文法结构、语系甚至文化差异。例如，英语与荷兰语同为印欧语系日耳曼语族，这两种语言间的机器翻译的结果通常比汉语与英语好。就科技汉英翻译而言，利用当前的主流机器翻译引擎有时可以得到可理解的翻译结果，但是词汇搭配和句式方面明显受到汉语文本的影响，表现出明显的翻译腔。对于语言错误的输入，机器翻译通常也无法分辨，只能将错就错，输出的译文自然千奇百怪，甚至完全无法理解。因此，目前想要得到准确通顺的汉译英翻译结果，机器翻译仍需要和人工编辑结合在一起，即机器翻译加译后编辑(MT+PE)，有时译前对待译文本进行适当的处理再进行机器翻译也是必要的。现在，也有一些科学家和业界人士在推动人机协同翻译，作为机器翻译完全取代人工翻译之前的过渡，通过在人工翻译过程中的人机交互提高翻译效率。很多计算机辅助翻译软件中已经允许加入机器翻译引擎提供译文，方便译员直接在计算机辅助翻译软件中进行机译编辑。

鉴于神经网络机器翻译强大的学习能力，相信在不久的将来，对质量要求不是太高的翻译工作完全可以由机器翻译处理，如电子邮件、网页的信息摘要以及基于计算机的信息服务等，机器翻译可能是唯一合适的解决方案。不过，目前机器翻译还没有达到可以自主完成翻译全过程的水平，不少学者甚至怀疑未来机器翻译也不一定能做到。在可预见的未来，人类译员的参与仍然是翻译过程中不可或缺的。现代翻译工具，包括机器翻译本身，也都是人类译员的辅助工具。专业翻译人员并没有因为机器翻译而真正失去工作，虽然使用机器翻译的人越来越多，但主要是为了了解文本的大致内容，而想要获得精确、正式的翻译，仍然需要求助于翻译机构或自由译者。

练习
practice

1. 简答题。
(1) 当前机器翻译的主要应用场景有哪些？
(2) 机器翻译的优点和缺点分别有哪些？

2. 利用两个以上机器翻译引擎翻译下列文字，对机器译文进行对比分析，然后搜索英文平行语料进行参照，对机器译文进行译后编辑，给出最终译文。

材料的硬度

人们可以通过多种方法测得材料的硬度。对于非金属材料而言，其中最简单的测试方法就是划痕测试法。如果某种物质 A 能够划破另一种物质 B，但 B 却不能划破 A，那么我们就说 A 比 B 硬。为方便起见，人们针对有代表性的矿物质统一规定了标准的硬度标度(其中，金刚石作为最硬的材料规定其硬度值为 10，滑石作为最软的材料其硬度值为 1)。

金刚石(C)	10	磷灰石($Ca_5(PO_4)_3F$)	5
刚玉(Al_2O_3)	9	萤石(CaF_2)	4
黄玉($Al_2SiO_4F_2$)	8	方解石($CaCO_3$)	3
石英(SiO_2)	7	石膏($CaSO_4 \cdot 2H_2O$)	2
正长石($KAlSi_3O_8$)	6	滑石($3MgO \cdot 4SiO_2 \cdot H_2O$)	1

目前，人们非常关注高硬度材料的研究与开发，如作为薄膜使用的透镜耐划痕性涂层等。在实践中，人们普遍感到金刚石与刚玉之间的标度差值容易引起误解，因为金刚石要比刚玉硬很多倍。对此，有人建议将金刚石的硬度记为 15，这样 9 和 15 之间的空白区间就可以分配给一些人工合成材料，如碳(C)和硼(B)的化合物。

硬度的现代标度体系(如 VHN 标度)是根据压痕硬度试验方法建立起来的。在压痕试验方法中，将压头施于被测材料的表面，然后通过测量压痕的尺寸大小来确定相应的硬度值。

第八章　科技翻译词汇处理

科技文体泛指所有涉及科学和技术的书面语及口语，其中包括科技学术著作、科技论文和研究报告、实验报告、各类科技情报、产品专利说明书和文字资料以及有关科技问题的会谈、会议、交谈用语、科技影片、录像等有声资料。科技英语指自然科学和工程技术领域的研究报告、科学著作、科学论文、科技教科书和科技演讲中广泛使用的英语。与文学类文本不同，科技文本不以语言的艺术美为追求目标，而是讲究逻辑的条理清楚和叙述的准确严密。科技文本最突出的功能是信息功能，主旨是传达科技信息、传播科技知识、进行科技交流。能够有效地、符合逻辑地向英语读者传递汉语科技文献承载的信息是每一个科技翻译人员从事汉英翻译的目标(任朝迎, 2017)。因此，研究科技文本翻译的重要性不言而喻。

科技文本翻译的一般原则如下：

(1) 信息准确真实。需要采用客观自然，符合科技规范标准的语言使读者获取正确的科技信息。

(2) 确保译文的可读性。考虑到中英语言差异，在翻译时不必拘泥于原文的形式，要做适当的灵活处理使得译文通俗易懂。

科技英语的词汇特征主要体现在以下几个方面：

(1) 含有大量专业术语。例如，在石油开采领域，lens 一词不是指常见的"透镜"，而应该翻译为"扁平矿体"。

(2) 常使用缩略语、符号、公式、图表表示科学概念。

(3) 大量使用名词化结构短语表达抽象的逻辑性。

科技文本的特点是语言简洁、逻辑鲜明，并且建立在一定的专业背景之上。没有扎实的科技知识背景，就无法读懂原文，更别提翻译了。此外，由于中西方思维的差异，英汉语言表现语义逻辑的方式截然不同。汉语强调悟性，重意义组合而轻形式结构，语法意义和逻辑联系常隐含在字里行间，要正确理解句义和语意必须从词语的意义、功能甚至语段、篇章和语境中去分析、寻找、体味；英语强调理性，重形式结构协调，以形显义、以形统意，注重语法关系和语义逻辑，要正确理解句义和语意必须分析词的形式、句子的结构和前后的逻辑关系(连淑能，2006)。因此，在做科技汉英翻译时，我们需要考虑如何在准确理解原意的基础上，通过必要的翻译手段，在保证科技文本客观性和专业性的条件下有效地实现科技信息传播。

国内学术界对科技英汉翻译的研究颇丰，但对科技汉英翻译的研究相对较少。本章基于汉语科技文本的词汇英译，从专有名词、准确性、灵活性、多样性和学术性五个方面探讨科技词汇英译的常见表现和处理方法。

第一节　专有名词

专有名词(proper nouns)是名词中比较独特的一类，它是相对于普通名词而言的，有的学者将其简称为"专名"，一般表示的是世界上独一无二的对象，在对人、地进行指称时具有唯一性，即所指对象是单一的、固定的，它的内涵往往仅仅依附于一个单独的事物(付吟璐，2012)。在科技翻译的时候，术语和专有名词非常重要。术语指某学科的专门用语，专有名词指某个(些)人、地方、机构、组织等专有的名称。这两类词汇涉及垂直领域核心概念，所以准确翻译科技文本特有词汇的内涵是对译员的基本要求。

科技文体承载着探索自然奥秘、揭示客观事物发展规律的功能，所传递的是客观真理和客观事实，因此，专业术语必须表意准确。不同领域的科技术语语义具有明确的层次结构，简明扼要，相对固定，具有国际通用性(韦孟芬，2014)。中国文化与西方文化差异较大，如语言差异、生活差异、宗教差异等，因此专有名词的翻译要跨越种种文化障碍。在翻译专有名词时，要讲究文化含量，注重意蕴，使翻译更符合目标语言的文化习惯，读起来才能朗朗上口。科技文本具有较强的学术性和专业性，所以更加讲究专业名词翻译的规范性和连续性。

为了准确传达源文本的科技信息，翻译科技专有名词时采用的办法如下：

(1) 正确运用各类查词工具(如权威词典、书籍、平行文本、在线术语网站等)获取约定俗成的既定译文，保证译文的专业性和准确性。

(2) 合理运用翻译手段使复杂的科技名词通俗易懂，如直译法(obstacle free zone 无障碍区)、形译法(U-bend 马蹄弯头)、音译法(nanometer 纳米)以及直译和意译相结合等。

(3) 保持积极主动的学习心态。科技不断发展会衍生出越来越多的新词汇，需要不断吸收新表达。

【经典赏析】

原文：华为轮值 CEO 徐直军表示："5G 移动网络可以实现逼真的视频通信体验。"

译文："5G mobile networks can provide a true-to-life video communications experience," said Eric Xu, rotating chief executive officer of Huawei.

赏析：本句在词汇上包含多种类型的专有名词，即机构、人名和科技术语。首先，华为技术有限公司(HUAWEI TECHNOLOGIES CO. LTD)是一家生产销售通信设备的民营通信科技公司，可简称为"华为"。该公司作为国内数一数二的企业，其重要地位不言而喻。翻译为 Huawei 是典型的音译法，与其官网翻译一致。"徐直军"作为人名，翻译时通常采用音译法，但考虑到他属于名人，一般在网络平台上都有自己的真实外文名或者既定翻译，可通过查询使得译文更加地道流畅。CEO 作为缩略语，是术语的一种，通过查阅资料我们可以知道该词意为"首席执行官"，也可直接翻译为 CEO，默认读者对常规缩略语的熟知。"5G 移动网络"和"视频通信"属是典型常规的科技术语，通过查找资料我们可直接得到既定译文，符合科技文本的客观性。本例句综合了科技专有名词翻译的特性，译文准确无误，自然流畅。

【案例解析】

◇ **例 1**

原文：正在实验的材料从放射性衰变中取得的电能是温插电材料从放射性衰变中取得电能的 20 倍。

原译：The materials they are testing would extract up to 20 times more power from radioactive disintegration than thermoelectric materials do from that.

解析：本句中的"放射性衰变"属于核医学名词，通过查找术语在线网站可找到对应的既定翻译，即 radioactive decay，该翻译采用直译法，读者可以清楚地明白该术语的内涵。原译将其翻译为 radioactive disintegration，未能保证术语翻译的正确性，属于误译。

译文：The materials they are testing would extract up to 20 times more power from radioactive decay than thermoelectric materials do from that.

◇ **例 2**

原文：法拉第当时不能计算出电磁波的传播速度，因为要计算传播速度这一任务需要用到麦克斯韦尔数学精度，而法拉第是根本不具备的。

原译：Faraday was unable to calculate the speed of propagation of electromagnetic waves, a task which required the mathematical precision of Maxwell, which Faraday entirely lacked.

解析：本句中涉及人名、科技名词和公式名称。人名翻译一般采用音译法，名人的名字已有既定翻译可借鉴，数学公式采用人称后置译法。原译把"传播速度"中的 "速度"一词翻译为 speed。考虑到术语的客观性和准确性，经过查证，"速度"翻译为 velocity 更具学术性和专业性，意为 the speed of sth. in a particular direction。

译文：Faraday was unable to calculate the velocity of propagation of electromagnetic waves, a task which required the mathematical precision of Maxwell, which Faraday entirely lacked.

◇ 例 3

原文：据中国载人航天工程办公室介绍，2 月 10 日 0 时 16 分，神舟十五号航天员费俊龙、邓清明、张陆密切协同，经过约 7 小时的出舱活动，圆满完成了梦天舱外扩展泵组安装等多项任务。

原译：The Shenzhou-15 astronauts on board the orbiting Chinese Tiangong space station completed their first spacewalk at 12:16 a.m. (Beijing Time) on Friday, according to the Office of China Manned Space Project. During the Extravehicular Activities (EVAs) lasting about seven hours, they completed several tasks, including the installation of the extension pumps outside the Mengtian lab module.

解析：本句中的"中国载人航天工程办公室"和"出舱活动"作为专有名词已有既定译文，需认真查证，切不可胡乱翻译。原译中将"载人航天工程"直译为 Office of China Manned Space Project，属于对专有机构名称的错译。"中国载人航天工程办公室"的官方译名为 China Manned Space Engineering Office (CMSEO)。"神舟十五号"和"梦天舱"均采用音译法，因为这是我国特有的科技产品名称，音译最为直接。对于"航天员"一词，通常想到的是 astronauts，但该词常见于描述美国航天员，中国航天员应该用 taikonauts 更为准确。

译文：The Shenzhou-15 taikonauts on board the orbiting Chinese Tiangong space station completed their first spacewalk at 12:16 a.m. (Beijing Time) on Friday, according to the China Manned Space Engineering Office. During the Extravehicular Activities (EVAs) lasting about seven hours, they completed several tasks, including the installation of the extension pumps outside the Mengtian lab module.

练习
practice

1. 简答题。

(1) 什么是科技专有名词？

(2) 科技专有名词的翻译方法有哪些？试结合实例论证。

2. 结合本节内容，翻译下列科技专有名词。

(1) 夏比冲击试验；

(2) 维氏硬度；

(3) AD；

(4) 蠕变；

(5) 延伸率；

(6) BBS；

(7) 算法；

(8) 云计算；

(9) 通信工程学院；

(10) 中国电子学会。

第二节　准　确　性

　　翻译时要保证用词的准确性指的是根据使用场合选用确切的词，选择恰当的词汇并且尊重英语词语的搭配习惯把原文的意思恰如其分地反映出来。英语中很多词是多义词，这给正确选词带来了一定的难度。译者不但要熟悉原文所涉及的专业技术知识，而且还要通晓词的基本含义和引申含义，以及词在特定科技领域中的特定含义(于建平，2001)。考虑到中英文在词汇风格上的巨大差异以及一词多义现象，再加上科技翻译对准确传达原文科技信息的基本要求，翻译科技词汇的准确性就非常重要。若不注重汉语词汇的深层次内涵以及英语词汇的多样性，就会导致词汇的滥用，造成误解，影响科技信息的有效交流。

　　科技翻译必须忠实于原文的内容，也就是说必须保持原文内容与译文内容之间的对等关系。科技翻译通常采用直译，实在不能直译时才采取意译。这就要求译者要熟练掌握原文和译文两种语言结构上的异同点，实施准确的语言转换。汉语中的同一词汇可能对应英文中的不同单词，译者需要结合语境和单词内涵准确辨别。如同为"侵蚀"，"土壤侵蚀"应该翻译为 soil erosion，而"金属侵蚀"则为 metal corrosion；科技文本中不可忽略图表的翻译，几何图形为 figure，简图和草图等为 diagram，而外形轮廓图为 profile。

　　翻译时，除了需要通过词汇的联立关系来确定多义词的词义之外，还需要根据语言的搭配习惯来处理相关词语(傅勇林，唐跃勤，2012)。汉语中的一致描述可能会对应英语中的不同搭配。尽管从语义理解上来说并不会造成阅读障碍，但科技文本对专业性和客观准确性要求高，在遣词造句上需要符合英语的习惯搭配，在翻译时需要使用更加地道准确的搭配，如密度"小"应该翻译为 low density 而不是 small density；"上税"的"上"应该翻译为 pay，而"上膏药"的"上"应该翻译为 apply。因此，在科技词汇英译时，要注意词义辨析，尊重英语搭配习惯，确保用词的准确性。

【经典赏析】

原文：承包商提出的设计变更以及对材料、设备的更换，需经工程师同意。

译文：The design change and replacement of materials and equipment proposed by the contractor shall be agreed by the engineer.

赏析：本句中的"变更"和"更换"在汉语理解上存在一定的包容性，都有"改变"的意思，所以译者会习惯性地将其翻译为 change。结合整体语境，"设计变更"有可能是对原来设计的简单改变，原本设计发生了变化，也有可能是设计得到了改进，所以用 change 比较合适；而"材料、设备的更换"指的是把 A 材料或设备替换成 B 材料或设备，强调的是一种"替换"，此时再用 change 不仅重复，且在具体含义上也有细微差别，因此译文采用了最为贴近原文内涵的 replacement。

【案例解析】

◇ 例 1

原文：实验首先展示了该手机的产品详情，页面包含名称、图片、价格等信息。

原译：The experiment showed the product details page of the mobile phone, including the name, picture and price.

解析：原译采用直译法，按照汉语语序，将中文文本与英文翻译一一对应，并且灵活地省略了中式语言表达中的习惯，即省略了"等信息"的翻译，译文可圈可点，看似无不妥之处。但仔细观察，原文中的"名称"一词是统称，实际上指的是该手机产品的品牌名称，而不能简单地翻译为

name。尽管这样的翻译并不影响读者的理解，但准确传达原文的科技信息是译者应有之义，在选词上需要更加贴切，因此翻译为 brand 更为准确。

译文：The experiment showed the product details page of the mobile phone, including the brand, picture and price.

◇ 例2

原文：通过文本识别对专利申请人进行划分，将每条专利的第一申请人划分出个人、企业、大专院校、科研单位四个类别，作为对应专利的申请主体。

原文：Through text identification, the patent applicant is divided into four categories: individual, enterprise, university and scientific research unit.

解析：中文重意合，在语言表达上讲究语义的连贯与逻辑鲜明。原文中的"通过文本识别对专利申请人进行划分"与后面的"作为对应专利的申请主体"存在一定的因果逻辑关系，读者阅读原文时可以把"作为对应专利的申请主体"作为补充说明，因此在翻译时可以采取适当的省略，将不言而喻的内容省译。因此，原译在整体上对原文本做了信息整合与适当省略，略去翻译"作为对应专利的申请主体"，体现出一定的翻译水平。但"大专院校"这一词选用 university，着实不妥。首先，中文中的"大专院校"是大学本科和专科院校的统称，university 通常可理解为"大学"。但美式英语的语境里它尤指同时包括了本科和硕博两个部分，且学生较多、规模较大的综合性大学。因此，原句中的"大专院校"用 college 更为准确，该词英文解析为 an institution where students study after they have left secondary school，在英语世界中的使用也更为普遍。

译文：Through text identification, the patent applicant is divided into four categories: individual, enterprise, college and scientific research unit.

◇ 例3

原文：表 5 中，街道总长度采取调研中常见值 70m。

原译：In Table 5, the total length of the street is 70m, which is common in the investigation.

解析：本句翻译的存疑点主要针对"调研"一词，原译选用 investigation 一词。经过查证，investigation 侧重于"审查"，牛津词典的解释为 an official examination of the facts about a situation, crime, etc，放在此处略有不妥，而 research 一词尤指大学或科研机构进行的研究和探讨。因此，结合本句语境，此处的"调研"一词用 research 更为准确。

译文：In Table 5, the total length of the street is 70m, which is common in the research.

◇ 例4

原文：从鱼群到蚁群，动物都表现出非凡的集群能力。

原译：From schools of fish to flocks of ants, animals exhibit extraordinary collective behavior.

解析：在汉英词汇翻译时，不仅要考虑单词的准确选取，也要尊重英文搭配习惯，使翻译更加地道。英文中对量词的描述要比中文丰富，如"一群"这个量词，我们可以用来搭配"一群人""一群牛"和"一群羊"等多个名词，但对应的英语却使用不同的单词，即 a group of people, a herd of cattle 和 a flock of sheep。因此，本句中"鱼群"和"蚁群"的量词搭配要准确，原译把"蚁群"翻译为 flocks of ants 属于英语量词的误用，flock 的英文解释为 a group of sheep, goats or birds of the same type，用来形容蚂蚁不太贴切，应改为习惯搭配 swarms of ants 更为准确。

译文：From schools of fish to swarms of ants, animals exhibit extraordinary collective behavior.

◇ 例5

原文：拆下或更换刀具时，请确保机器已冷却至安全温度，防止灼伤。

原译：When dismantling or replacing the knife, please ensure the machine is cooled to safe temperature to prevent burning.

解析：原译把"刀具"翻译为 knife 有些片面。通过搜索网络，"刀具"一词指的是一种日常生活常用的工具，亦可是一种武器；在机械制造中用于切削加工的工具称为切削工具，也可以理解为金属切削刀具。该词的专业译名为 tool，原译使用的 knife 一词可解释为 is a tool for cutting or a weapon and consists of a flat piece of metal with a sharp edge on the end of a handle，结合上下文语境，用在此处不妥。因此，在翻译科技文本时，首先要正确理解原文，再通过搜集资料查找相关译文表达加以证明，以免以偏概全，不能准确表达原意。与"刀"相关的类似科技表达还有刀架(tool rest)、刀片(tool blade)以及刀座(tool apron)等。

译文：When dismantling or replacing the tool, please ensure the machine is cooled to safe temperature to prevent burning.

练习
practice

1. 简答题。

(1) 如何理解科技词汇英译的准确性？

(2) 结合实例分析科技词汇英译时用词不准确的现象，并加以改正。

2. 结合本节内容，翻译下列句子，特别注意划线词翻译的准确性。

(1) 如果发放的配额少于实际排放量，企业会在碳交易市场上<u>购买</u>配额，以此满足排放需求。

(2) 在高速铁路向中西部延伸的过程中，沿线地质条件越来越复杂，<u>工程难度</u>不断增加。

(3) 银行客户保密指的是银行对涉及<u>客户</u>的所有事实的保密义务。

第三节　灵　活　性

翻译的灵活性是处理不同语言转换的一条关键原则。一切翻译都应该尊重语言本身的差异，结合目标语读者的语言习惯，不必拘泥于原文的语言框架，要在合理范围之内进行自由转换。尤金·奈达曾指出，"语言之所以互不相同，主要是因为它们具有互不相同的形式，因此翻译中如要保存原作的内容，就必须改变表现形式"(谭载喜，1984)。不仅科技文本翻译要做到在准确表达原文含义的基础上使译文通顺流畅、避免翻译腔，所有的翻译都应该是"语言的再创作"，而并非简单的"双语转换"。

本节讨论的科技词汇翻译的灵活性主要体现在以下三个方面。

(1) 词汇翻译时的增译与省译现象。在科技翻译中，由于汉英两种语言的使用习惯不同，不增一词或不减一词的直译有时行不通，所以，常常在保持原有意义不变的基础上适当地增加一些词组或句子，使译文结构完整，表达流畅。增译指根据原文上下文的意思、逻辑关系以及译文语言的句法特点和表达习惯，在翻译时增加原文字面没有出现但实际内容已出现的词。英译的增译体现在增加冠词、代词、名词、介词、连词、概括性词等多个方面。例如，"我是建筑师"翻译为 I'm an architect, 增加了冠词 an；"有些方法是直接的，有些是间接的"翻译为 Some methods are direct, but others are indirect，增加了连词 but。省译指删去不符合目标语思维习惯、语言习惯和表达方式的词，以避免译文累赘。由于英汉两种语言的表达习惯不同，翻译过程中一字不漏地照搬会使译文变得累赘，不符合行文习惯，严重的还会使文章内容产生歧义，因此适当地运用省译法可以使译文言简意赅地表达清楚。英译的省译体现在省略中文中常见的范畴词或者一些不影响原文理解的次要成分，例如，"感应现象"翻译为 induction 而不是 induction phenomenon；"修改方案"翻译为 modification 而不是 modification plan，等。

(2) 词汇翻译时的词性转换现象。词性转译法是汉译英中常使用的翻译方法，指根据汉语的规范表达方式进行词性转换。例如，由于汉语多用动词，因此在做汉译英时，时常将汉语的动词译为英语中的名词、形容词、副词、介词短语等。科技英语中常常使用名词化结构来体现文本的客观性和学术性，而汉语的语言习惯倾向于大量使用动词，因此在做科技汉英翻译时，可以采用灵活的处理方式，对翻译词汇的词性进行转换，使之符合英文行文习惯，体现出翻译的灵活性。翻译的最终目的是要达到与原文内容信息和功能上的对等，而不是追求形式上的完全对应，所以为了避免翻译实践中出现片面追求形式上的对应，可以选择使用词性转换法以表层结构形式的偏离换取内容或信息的一致。翻译时可以将汉语的动词转译为名词、动词或者介词等，例如，"返回地球是个大问题"可译为 The return to earth is a big problem；"它能在水中溶解"译为 It is soluble in water；"实验结束了"译为 The experiment is over。

(3) 词汇翻译时的可编辑性。这里强调的是译者根据已有知识或者逻辑常识对原文中一些明显的错误进行判断，通过求证原文作者对原文进行修改后再翻译。例如，"2022 年 2 月 30 日，XXX 医院接待了……"，原文很明显出现了错误，2 月并没有 30 号，所以译者绝不可机械地翻译为 On February 30th，需要求证原文的真实性，灵活处理。

【经典赏析】

原文：每个省级行政单位在每个年份的 Y02 专利申请总数是固定的，多数专利只含有一个 Y02 分类号，但仍然有一定比例的专利含有两个及以上的 Y02 分类号。

译文：The total number of Y02 patents applied every year in each province is fixed. Most patents have only one Y02 classification number, but some patents have two or even more Y02 classification numbers.

赏析：原文中的"每个省级行政单位"，其实指的就是"各个省"，"有一定比例"其实就可以简单地认为表示"一些"。因此在翻译时，无需拘泥于原文框架，将之翻译为 provincial administrative unit 以及 have a certain proportion，这样的直译未能做到简洁明了，具有很浓厚的中式英语色彩。译文对原文表达进行了灵活处理，省略了不必要的表述，言简意赅地翻译为 each province 以及 some，体现出译者良好的翻译水平。

【案例解析】

◇ 例 1

原文：因为铸件冷却时收缩，所以铸模应该大一些。

原译：The mould should be a little larger as the castings shrink as they cool.

解析：汉语往往只说出中心词就可以把意思表达清楚，而英译时必须补全该中心词的修饰语。例如，"不同的物质"其实指的是"不同种类的物质"，"热处理"指的是"热处理过程"，"500℃左右"指的是"温度 500 度左右"。面对类似的汉语表达，需要利用介词 of 作后置定语，并在前面加上表示度量等意义的名词。原文中的"铸模"其实表示的是"铸模的尺寸"，但原译只是机械的翻译，并没有灵活处理，适当补充，此处翻译为 the size of the mould 更为贴切。

译文：The size of the mould should be a little larger as the castings shrink as they cool.

◇ 例 2

原文：在制作酸奶的过程中，达能发现了一种酶可以使鲜奶保鲜长达 4 小时。

原译：In the process of developing the yogurt, Danone discovered an enzyme that preserves fresh milk for up to four hours.

解析：汉语中常见的一些中式表达，例如，"在 XX 过程中""在 XX 进程中"以及"在 XX 下"等。面对这样的表述要灵活处理，翻译时要适当省译中文的范畴词，直接表示动作。原译处理为 in the process of developing…，过于直译，显得啰嗦，可以直接处理为 developing…，再加上连词 while，使译文通顺流畅。

译文：While developing the yogurt, Danone discovered an enzyme that preserves fresh milk for up to four hours.

◇ 例 3

原文：有关荷兰最新发明的报道，使伽利略改变了个人工作的进程。这项发明是一台新的光学仪器，它能使远处的物体看起来好像在近处似的。

原译：Galileo changed the course of his private work for reports of the recent invention in Holland. It was a new optical instrument that made distant objects appear close.

解析：原文中的"改变"一词原为动词，翻译时可处理为"名词形式的 change"。考虑到科技文本的客观性以及科技英语常用名词性短语，此处可以进行词性转换，将动词改为名词译为 the change of…，更为地道、客观，弱化人为色彩。

译文：The change in the course of Galileo's private work was caused by reports of the recent invention, in Holland, of a new optical instrument that made distant objects appear close.

◇ 例4

原文：由于大多数金属具有韧性和延展性，所以它们可以压成薄板和拉成钢丝。

原译：As most metals have properties of malleable and ductile, they can be beaten into plates and drawn into wire.

解析：在说明事物的性状时，汉语习惯用名词，例如由"度""性"等词构成的名词"强度""硬度""密度""弹性""流动性"等构成主谓短语来表示(往往用形容词作谓语：强度大、流动性好等)，但科技英语却往往习惯用表示特征的形容词及其比较级作表语(严俊仁，2010)。例如，XXX is strong/ is fluid。原文中的"韧性"和"延展性"可直接翻译为形容词，无需单独翻译出"性"，避免翻译腔，采用灵活的方式进行翻译词性转换。

译文：As most metals are malleable and ductile, they can be beaten into plates and drawn into wire.

◇ 例5

原文：如何重塑地下商业街转交空间形态以提升地下环境的整体品质，是建筑师地下空间设计时需要着重考虑的因素。

原译：Therefore, how to reshape the form of the transfer space of the underground commercial streets to improve the overall quality of the environment is a factor that architects need to focus on when designing underground space.

解析：原文的"转交空间形态"会让人一头雾水，不知所云。经查证，原文中的"转交"是"转角"一词的误写。原译并未识别出这个基本错误，而顺着原文进行翻译，这对准确传达原文科技信息造成极大困难。这启示我们，在翻译时，既要忠实于原文，又要具备一定的翻译素养。针对这类基本错误，译者要有能力识别并积极求证，切莫照搬照抄，胡乱翻译。此处的"转交"应该为"转角"，翻译为 corner space。

译文：Therefore, how to reshape the form of the corner space of the underground commercial streets to improve the overall quality of the environment is a factor that architects need to focus on when designing underground space.

练习 practice

1. 简答题。
(1) 科技词汇翻译的灵活性体现在哪些方面？
(2) 除了本节提到的灵活性的三点表现，你还能想到哪些？结合实例证明。
2. 结合本节内容，从词汇翻译灵活性的角度赏析下列句子的翻译。
(1) 由于利润扩大，新产品不断推出，投资者认为没有必要进一步对抗。
With profits and new products blooming, few investors see a need for further confrontation.
(2) 喜忧参半的经济信号让人怀疑这个国家需要多长时间才能走出衰退。
Mixed economic signals call into question how long it will take for the country to emerge from recession.
(3) 人类目前深为担忧的是，在不太遥远的将来，可能会出现煤、石油、天然气或其他燃料来源紧缺的现象。
What is worrying the world greatly now is possible shortage of coal, oil, natural gas or other sources of fuel in the not too distant future.

第四节　多　样　性

科技英语是现代英语中一个概况性的功能变体名称，其作用是传播自然科学或社会科学的知识和技术。在科学技术的发展过程中，科技英语已形成了一种独立的文体形式。科技文本自身具有较强的学术性，正确理解原文需要一定的专业知识背景，而且科技文本逻辑鲜明，在翻译为英语时常使用被动语态或者长难句。中英双语之间本身就存在巨大差异，中文重视重复，起到加强语气强调的效果，而英文尽量避免重复，强调词汇的多样性和丰富性。一般来说，除非有意强调或出于修辞的需要，英语总的倾向是避免重复。尽管科技文本的文体风格较为单调，个人色彩很轻，大多使用专有名词翻译，也有很多习惯表达。但在汉英翻译时，一味地重复相关词汇或者使用相同的句子结构，往往会让英语读者感到厌烦。因此，翻译时可以在一定范围内使用多种表达来避免重复，增强趣味性和可读性。

　　本节所强调的"多样性"主要指的是选词的多样性和词汇表达的多样性。选词的多样性与"词汇的准确性"关系密切，多样性可以是在词义相同的情况下选择同义词进行替换，避免重复，但并不代表不考虑语境和具体内涵一味地选择替换词，这样做可能会造成理解偏差。例如，"XXX 认为……"既可以翻译为 XXX stated…，也可以翻译为 XXX suggested…以及 XXX said…等。词汇的多样性还体现在表达上，比如，科技文本中常见的"在这种情况下"这一表述就有多种译法，如 under these/such circumstances，such being the case，in accordance with the specific conditions 等。从事科技汉英翻译时，要充分体现词汇的多样性，避免重复，符合英语表达的风格。

【经典赏析】

原文：这一发现可以帮助人们了解宇宙更奇怪的一面——暗能量。

译文：The finding may help shed light on an even more bizarre aspect of the universe—dark energy.

赏析：译文将"帮助人们了解"翻译为 help shed light on 更加生动形象，强调了"了解"这个动作，而且 shed light on 一词本身就是"阐明、使……清楚地显示出"的意思，更加贴合语境，也可以体现科技文本的学术性。如果使用 know 一词，其本身也有 to realize, understand or be aware of sth. 的意思，或者使用 comprehend、learn about、understand 等词语，虽然在整体理解上并无差异，但不如 shed light on 更具体、更客观。结合之前提到的"词汇的准确性"以及"词汇的学术性"，译文的处理更地道、更准确。这也启示我们，科技翻译词汇多样性的使用一定要基于正确的词义和得体的风格，在适当范围内变换。

【案例解析】

◇ 例 1

原文：调查的结果显示，吸毒不仅仅发生在大城市。

原译：Results of the survey showed that drug use is not confined to our major cities.

解析：原文中的"显示"一词是科技文本中常见的表述，例如，"实验结果显示……""研究报告显示……"以及"图 1 显示……"等。这一词也可替换为"说明""表明"以及"表示"等，因此该词语通常可翻译为 show、 indicate、express 或者调整句式翻译为 it is shown that…等，翻译方法多样。原译将之翻译为 showed，语言色彩较为平淡，可换为 highlighted the fact that…，突出调查结果，强调一种客观事实。

译文：Results of the survey/investigation highlighted the fact that drug use is not confined to our major cities.

◇ 例2

原文：一个更复杂的问题是宇宙中的暗能量是否在变化。

原译：A more complicated question is whether the amount of dark energy in the universe is changing.

解析：原文中 "复杂" 一词在英语中可找到多个单词翻译，例如 complicated、intricate 和 complex 等，在传达原意上没有什么影响。complicated 的英文释义为 made of many different things or parts that are connected; difficult to understand，放在此处也能翻译通顺。但结合语境，此处的 "复杂" 可以理解为 "让人头疼的、难懂的、难解的、难处理的"，因此可以用 awkward 来翻译，其英文释义为 difficult to deal with。

译文：A more awkward question is whether the amount of dark energy in the universe is changing.

◇ 例3

原文：手机可以作为公众获取重要信息的另一种途径。

原译：The mobile phone can serve as another way for the public to access important information.

解析：原文中的 "途径" 一词可以理解为 "方法" "手段" "方式" 等，英语中的同义替换词有很多。为了避免重复，翻译时可以选用多个词进行替换，但要注意不同单词的介词搭配，例如，a method of…、a way of…以及 an approach to…等。译文选用 avenue 一词，英文释义为 a choice or way of making progress toward sth.，更加生动形象，增强了词汇的丰富性。

译文：The mobile phone can serve as another avenue for the public to access important information.

◇ 例4

原文：毫无疑问，牛奶产量已降至令人惊讶的极低水平。

原译：There is no doubt that milk production has fallen to a surprisingly low level.

解析：一般来说，见到"毫无疑问"，我们都会把它翻译为 There is no doubt that…，这是英语学习者的一种固定思维，把耳熟能详的英文表达牢记于心。若在翻译时一味地采用重复表达，会显得非常累赘。因此，我们可以灵活采用多种表达，增强语言魅力，学习积累更多英语搭配。例如，该短语也可以翻译为 It is indisputable that…，Undoubtedly，It was without a shadow of a doubt，There is little dispute that…以及 sth. showed beyond doubt that…等。

译文：There is little dispute that milk production has fallen to a surprisingly low level.

◇ 例 5

原文：大量证据表明，维生素 D 和钙有助于治疗骨质疏松。

原译：There is plenty of evidence that vitamin D and calcium help treat bone loss.

解析：科技文中常见的"研究表明"或者"证据表明"这类表述，翻译时常常会使用 There be 句型或者 It 句型。与本节提到的例 1 类似，我们不仅可以自由变换"表明"一词的翻译，也可以把目光放在"大量证据"的翻译上，让物作主语，灵活翻译。

译文：Considerable evidence has accumulated that vitamin D and calcium help to treat bone loss.

练习
practice

1. 简答题。

(1) 如何理解科技汉英翻译中词汇的多样性？

(2) 你认为掌握翻译的多样性对语言学习有什么好处？结合自身实际分析。

2. 结合本节内容，从词汇翻译多样性的角度分别给出下列句子的多种英译版本，尤其注意划线词的翻译。

(1) 我们现在取得了巨大进展，许多国家的疟疾病例显著减少。

(2) 最终的解决方案是做广告，并且毫不夸张地说，谷歌现在本质上是一家广告公司，因为这几乎是它所有收入的来源。

(3) 杜克大学的特里·莫菲特和她的科研同事们发现，有自制力问题的孩子在长大成人后往往会面临一系列更棘手的问题。

第五节　学　术　性

　　科技文章一般平铺直叙，力求平易和精确，较少使用修辞手段，避免行文晦涩、表露个人感情和论证上的主观随意性，因而读起来总给人一种庄重甚至是压抑的感觉，这是由科技文章的文体特点所决定的。科技文本与日常文本或者文学文本有所不同，其不具备较强的感性形象思维，感情色彩单一，常使用第三人称以此突出科技文本的客观性与逻辑性。科技语言在形式上的特点就是尽量避免使用第一人称代词，大量使用无主句和物主句来表达客观性，科技英语在反映客观性要求方面已经形成一些惯用句式，如被动式、物主句等(郊春生，2008)。在遣词造句上，科技文本尽量避免使用口语化的词语和短语，多采用正式规范的书面词语来替代，语言表达上注重行文简洁明了、表述客观流畅。

　　众所周知，科技英语涉及科学技术，科技人员在研究和解决科学技术问题时，必须要从客观事物出发，尽可能作出客观、准确的描述和论证。为了保证这种客观性，从语法结构上讲，科技文献自然会采用大量以客观事物、研究对象为句子主语的句式非人称句，采用表示客观事实的时态等等(冯志杰，1999)。因此，在做科技汉英翻译时，要尊重科技文本的学术性和客观性，避免使用口语化表达。可以采用被动句、名词性结构、一般现在时，选用正式词汇、常见科技表达等多种方法体现文本的学术性。例如，科技文本中常见的"XXX 认为……"可以翻译为 XXX claims that…而不是 XXX thinks that…；"研究发现……"可以翻译为 It is found that…或者 Among its findings is the fact that…；科技文本中常见的"下文中……"，应该翻译为 in what follows 而不是 in the following paragraph。结合科技英语本身的特点，比如大量使用被动句、名词化结构以增加客观色彩等，在英译科技文本词汇时要注意选词，必须能够体现原文的学术性，符合科技文体规范。

【经典赏析】

原文：大约 20%的人患有慢性肝病，其中许多人需要进行肝移植。

译文：Approximately 20% of these people develop chronic liver disease, and many require liver transplantation.

赏析：从整体上来说原文并无专业词汇，语言简洁，通俗易懂，译文在翻译上也保证了原文的流畅度，可圈可点之处就是译文对"大约"一词的处理。提到"大约"一词的翻译，首先想到的就是 about。"大约 50 人"翻译为 about 50 people，"大约 400 米"翻译为 about 400m，这无可厚非。但译文采用了 approximately 一词，《牛津词典》对该词的解释为 used to show that sth. is almost but not completely accurate or correct，从对原文的理解上来说，这两个词语都可以表示"大约"的内涵，词义相似但也存在细微差别。approximately 这个单词更能表达出"有多接近准确数字"的功能，且常用于正式语体，而 about 常用于口语或者一般性文体中。因此，本句的翻译用 approximately，更符合科技文体特征，更能体现学术性。

【案例解析】

◇ 例1

原文：这种评论间的对比是不公平的，因为基于累计帮助性投票的评论帮助性计算方法会不可避免地认为更早的评论更有帮助。

原译：The comparison is unfair, because the calculation method of comment helpfulness based on cumulative helpfulness voting will inevitably think that the earlier comments are more helpful.

解析：这里主要对"认为"一词展开讨论。英文中可以表示"认为"含义的单词有很多，比如 think、believe、consider 以及 regard as 等。think 一词使用频率最高，《牛津词典》对它的解释为 to have a particular idea or opinion about sth/sb; to believe sth，此处用该词翻译也未尝不可，并不会影响读者对原文的理解。但 think 一词使用范围太广，且更常见于口语化表述中，原文是对某种研究内容的客观性描述，应该采取更具学术性的词汇来表达。因此，选用 reckon 更符合科技文本的学术特征。

译文：The comparison is unfair, because the calculation method of comment helpfulness based on cumulative helpfulness voting will inevitably reckon that the earlier comments are more helpful.

◇ 例2

原文：下面将对这三个要素与心理感知的关联性进行研究。

原译：In the following paragraph, the correction between these three factors and psychological perception will be studied.

解析：原译采用直译，按照原文的语序结构进行翻译较为流畅。科技文本在描述研究过程和发现时，常常会使用"下面……""接下来……""下文中……"等逻辑衔接语。原译直译为 In the following paragraph，不够凝练和准确。In what follows 更能体现出客观性与学术性。

译文：In what follows, the correction between these three factors and psychological perception will be studied.

◇ 例3

原文：缩拢双唇快速吹气时，你会感到吹过手面的空气是凉的。

原译：When you close your lips and blow quickly, you will feel the air over your hands is cool.

解析：原译使用条件状语从句，突出了主语 you 的个人感受，但其实原文想要强调的是一个现象，即"缩拢双唇快速吹气可让人感觉吹过手面的风是凉的"。因此，在翻译时，应该突出条件"动作"，忽略受众"你"，使译文可以准确传达原文想要表述的现象。另外，可使用科技英语中常见的祈使句来使译文的动作描述更为客观，避免使用从句造成译文冗余，不够简洁明了。句首用祈使句表示条件或假设再由 and 引出主句表示该条件或假设引出的结果或推论，语气更加客观正式。

译文：Pucker up your lips and blow fast, and the air that passes over your hand feels cool.

◇ 例4

原文：在今后很长一段时间内，我们仍将把钢及其合金作为重要的工业材料.

原译：We will still regard steel and its alloys as leading industrial materials for a long time to come.

解析：由于中英文语言习惯的不同，中文常常会用"人们""大家""我们"等一类任指词语作为主语，而在翻译为英语时常用被动语态使句子精炼，且突出受动者，突出"物"，避免人作主语使得语言描述带有个人色彩。因此，翻译时可将本句中的主语"我们"省去，宾语"钢及其合金"变为主语，并使用科技英文中常见的被动语态使译文更客观，减少个人色彩对原文学术性的影响。

译文：Steel and its alloys will still be taken as the leading materials in industry for a long time to come.

◇ 例5

原文：这一假设可能并不准确，因为峰宽随着能量的增加而增加，但随机求和产生的损失似乎对同位素比测量没有实质性的影响。

原译：This assumption may not hold precisely because peak widths increase with energy, but it appears that the losses occurring from random summing don't materially affect isotopic ratio measurements.

解析：尽管，科技英语中的缩略语使用广泛，但在正式的科技英语写作中，要充分体现文章的学术性应避免使用加 not 缩写的错误表达，因为缩写是口语表达，如 it isn't、it doesn't、we didn't、it can't、hasn't、needn't 等。正确的表达是 not 与系动词不缩写。

译文：This assumption may not hold precisely because peak widths increase with energy, but it appears that the losses occurring from random summing do not materially affect isotopic ratio measurements.

练习
practice

1. 简答题。

(1) 如何定义科技汉英文本翻译时选词的学术性？

(2) 结合实例阐述科技文本学术性的必要性和重要性。

2. 结合本节内容翻译下列句子，体现用词的"学术性"，尤其注意划线词的翻译。

(1) 据美国人口统计局的数据，大约有 4 000 万 65 岁以上的老人独居。

(2) 属于这类物质的有云母、瓷器、石英、玻璃、木材，等等。

(3) 但是，目前大部分的研究集中于如何在教学活动中运用情感策略，而对于教材中如何体现情感因素却较少有人研究，理科教材尤为严重。

第九章　科技翻译句法处理

　　科技文体在表述科学原理、规律、概念以及事物之间错综复杂的关系时，需要客观地陈述事实和揭示真理，因此在逻辑上讲求条理清晰，思维上讲求准确严密，表述上讲求客观规范。而在英语和汉语中，科技文体的句式都比较复杂。本章将从科技文体的特征入手，探讨科技翻译句法处理的策略。

　　句子是一个完整的语义单位，所以现代翻译理论普遍认为句子是最重要的翻译单位。汉语重意，造句多用意合法，主要凭借隐含的逻辑将句子各个部分统一起来，所以许多情况下很难从汉语句子的外形上确定句子的主干和信息重心。而英语重形，造句多用形合法，以词汇为纽带，借助连词、关系代词等形式，使句子意义明确、结构紧凑。所以汉译英时只有正确地确立汉英句子的主干和信息重心才有可能译出地道的英语，而不会陷入误译的沼泽。

　　科技文本通常需要准确客观地传达信息，且有时一个句子可能包含多重信息，这就要求译者对原文整体把握并将长句合理划分，在保证语义完整的基础上，准确清晰地传达原文的信息与态度。受大众对科技文本的刻板印象影响，译者常青睐长难句，而本章将从具体实例说明灵活运用长短句式的必要性和重要性。另外，本章还将探讨科技翻译中如何实现语言简洁、表达清晰准确以及遵循句子统一性原则的基本思路。

　　总的来说，科技翻译的过程就是译者在用词、句式、语态以及表达等方面不断做出选择的过程，对各种翻译方法和翻译技巧的使用也不可一概而论，要具体问题具体分析，做出最优选择。

第一节　语言简洁

　　语言简洁是科技翻译的重要标准之一，这意味着在翻译科技文本时要避免冗长复杂的表达，避免造成理解困难。译文应突出句子主干，使用简洁恰当的修饰成分，这样更易于理解和有效传达概念。

　　表达有效意义的最小单位是句子，而句子的表达有很多规则，比如简洁、连贯、多样性等等。本节主要探讨科技文本中语言表达的简洁性，主要有四种策略：使用名词化结构代替从句，用分词短语、独立主格结构或 with 结构来代替从句，使用介词结构以及使用代词。

【经典赏析】

原文：自从互联网发明以后，我们不得不处理比以往多得多的信息。

译文：Since the invention of the Internet, we have to deal with much more information than ever before.

赏析："自从互联网发明以后"可以译为 since the Internet was invented，虽然没有语法问题，但略显啰嗦。译文使用名词 invention，句子结构变为简单句，表达更简洁。这里的简洁不一定指句子的长短，而是语法结构的简化。

(1) 使用名词化结构代替从句。名词化是语言学的重要概念。韩礼德(M. A. K. Halliday)(1976)从语境和交际功能出发，把名词化和语法隐喻联系起来，认为名词化是语法隐喻的主要来源，是概念隐喻的主要手段。名词化是将过程(其词汇语法层的一致式为动词)和特性(其一致式为形容词)经过隐喻化，以名词形式体现的参与者，而不再是小句中的过程或修饰语，从而提高语言的简洁性。科技英语中广泛使用名词化结构，如表示动作意义的名词＋of＋名词＋修饰语，这种结构往往起到从句的作用，可避免过多使用主谓结构，使行文简洁、表达客观、内容准确，而且信息量大。同时，这种结构强调存在的事实，而非行为本身。使用名词化结构代替从句，将从句转化为名词短语是一种简化科技翻译中复杂句子的方法。

【案例解析】

◇ 例 1

原文：考察一下音频带宽，我们便可以看出高频的重要性。

原译：We can see the importance of higher frequencies if we examine the bandwidth of the audio frequency.

解析：原译中的条件状语从句，可以用名词化结构来代替，使句子变为简单句。因此译文使用名词 examination 进行简化，同时，为体现科技文本语言的客观性，用被动句替代人称主语 we。

译文：The importance of higher frequencies can be seen by an examination of the bandwidth of the audio frequency.

◇ **例 2**

原文：旁路直流回路并入后，电气阻抗组成系数将发生变化。

原译：The composition coefficient of electrical impedance will change after a bypass dc circuit is connected.

解析：原译中的 after 引导的是时间状语从句，译文使用名词 connection 来简化状语从句，如此一来，句子变为"主谓"型的简单句。句子的长短，似乎没有变化，但简洁不仅仅是指句子长短方面，更重要的是语法结构。

译文：The composition coefficient of electrical impedance will change after the connection of a bypass dc circuit.

在科技翻译中，不仅可以使用名词化结构来实现语言简洁，还可以使用破折号巧妙地简化句子。

◇ **例 3**

原文：该系统有两种形式，分别是单输入单输出和多输入多输出模式。

原译：The system has two forms, which are single input single output and multiple input multiple output.

解析：原译用的是定语从句，译文使用一个简单的破折号替代了从句，句子的表达更简洁，表意也非常清晰。

译文：The system has two forms — single input single output and multiple input multiple output.

(2) 用分词短语、独立主格结构或 with 结构来代替从句。

◇ **例 4**

原文：电子可以在晶体结构中自由移动，这使它具有很好的导电性。

原译：The electrons can move freely in the crystal structure, which makes it have high electrical conductivity.

解析：原译含有 which 引导的定语从句，表达虽然没有问题，但显得比较僵硬。根据句意，"使它具有很好的导电性"，可以使用动词短语 contribute to，用分词结构来代替从句。这两个句子的主要区别在于语法结构，前者是定语从句，后者是带有分词结构的简单句。实际上，有研究表明，英美人写作时使用从句的比例并不多，他们更多使用的是语法结构简单，但其中含有比如分词、同位语、插入语等各种附加结构的简单句，这也是在写作中锻造句子的一个重点方向。

译文：The electrons can move freely in the crystal structure, contributing to its high electrical conductivity.

◇ 例 5

原文：训练网络包含 8 个特征提取块，每个特征提取块包含 3 个提取层。

原译：The training network contains 8 feature extraction blocks, and each feature extraction block contains 3 extraction layers.

解析：原译的表达有些烦琐，比如 and 前面的分句中包含 feature extraction blocks，后面又重复使用了该短语。要避免这种重复，不妨采取独立主格结构。译文中的 each containing 3 extraction layers 属于独立主格结构，它使句子变得更加简洁。

译文：The training network contains 8 feature extraction blocks, each containing 3 extraction layers.

◇ 例 6

原文：图 1 是基于 PSCAD 软件搭建的 2 车仿真模型，A、B 两车相距 1700 m。

原译：Fig. 1 is a 2-train simulation model based on the PSCAD software. The trains of A and B are 1, 700 meters apart.

解析：原译由两个简短的句子构成，而且句意非常紧密，这里的"A、B 两车相距 1700 m"完全没有必要用一个单独的句子来表达。只需要用一个 with 结构，句子表达不仅更简洁而且更紧凑。

译文：Fig. 1 is a 2-train simulation model based on the PSCAD software, with the trains being 1, 700 meters apart.

(3) 使用介词结构。前面探讨了使用名词化结构、分词短语等来改写句子，使句子表达更简洁。但表达简洁的方法多种多样，以下内容将聚焦于使用介词结构。

◇ 例 7

原文：由于频率较高，前视合成孔径雷达系统存在一定的模糊性。

原译：Because it has a high frequency, the forward-looking SAR system faces some ambiguity.

解析：对于"由于"，原译使用了 because 引导原因状语从句，实际上，可以通过介词短语来体现"由于"的意思，比如使用 due to，其后面用名词短语，句子由状语从句变为简单句，表达更简洁。这里的简洁主要在于句子结构而非句子的长短。如要表达"由于、因为"，除了使用 due to + 名词短语,还可以考虑 because of + 名词短语,owing to + 名词短语的结构。通过使用介词结构,可以使句子表达更简洁,同时,变换使用介词结构也能实现句子表达的多样性。

译文：Due to its high frequency, the forward-looking SAR system faces some ambiguity.

◇ 例8

原文：人工智能作为一项尖端技术,近年来发展迅猛。

原译：Artificial intelligence is a cutting-edge technology, and it has developed rapidly in recent years.

解析：原译是一个并列句,通过并列连词 and 连接前后两个分句。对于第一个分句,可以使用介词 as,将它变为一个更简单的结构,即 as a cutting-edge technology。这样的句型在英文中经常出现,比如,As an efficient means of communication, email has been well received in the workplace. (电子邮件作为一种有效的交流方式,在职场中受到大家的普遍欢迎。)

译文：As a cutting-edge technology, artificial intelligence has developed rapidly in recent years.

◇ 例9

原文：阻抗的减小有利于再生制动能量的吸收,这与理论分析是吻合的。

原译：The decrease of the electrical impedance is conducive to the absorption of regenerative braking energy, which is in agreement with theoretical analysis.

解析：原译用的是 which 引导定语从句,可以将定语从句改写为介词结构,也就是去掉 which is,从句就简化为介词结构 in agreement with theoretical analysis。如此一来,句子由之前的复杂句变为简单句,同时句子的长度也压缩了。因此,如果要表达"与理论分析是吻合的"(in agreement with theoretical analysis),"与实验结果是吻合的"(in agreement with experimental results),"与观察到的情况是一致的"(in agreement with observations)等等这样的概念,都可以使用这一句型。此外,为使译文表达多样,避免重复,"吻合"除了用 in agreement with,还可以用 be consistent with。

译文：The decrease of the electrical impedance is conducive to the absorption of regenerative braking energy, in agreement with theoretical analysis.

◇ 例 10

原文：对于已经运营的城轨交通工程，该方法不具备经济性。

原译：For the urban rail transit projects that have been put into operation, this method does not work economically.

解析：原译使用定语从句来表达"对于已经运营的城轨交通工程"。实际上，只需要一个介词短语 in operation，便可以表达"已经运营"的意思，这样句子就可以大幅简化。

译文：For the urban rail transit projects in operation, this method does not work economically.

(4) 使用代词。代词的使用在翻译中常容易被忽略，但它对简化句子其实大有帮助，还能避免重复。

◇ 例 11

原文：该方法使模型更轻便，提高了训练效率。

原译：This method makes the model lighter and improves the training efficiency of the model.

解析：原译没有语法问题，但仔细观察一下，可以发现 model 在句中重复了。通常在英文表达中，尤其在同一个句子中，最好不要出现重复现象，可更换为其他词，如使用代词、同位语等等。译文使用代词 its 代替了 the model，这样一来，and 前后的内容比较对称，句子表达更简洁。

译文：This method makes the model lighter and improves its training efficiency.

◇ 例 12

原文：为了确定试样的主要参数，我们对其进行了标记。

原译：In order to identify the main parameters of the test specimens by their labels, the specimens were all labeled.

解析：原译重复的内容有 specimens，还有 labels，实际上，这两个重复都可以避免。可将句子的主干结构放在前面，目的状语放在后面，主干结构中用名词，目的状语中用代词。这样一来句子变得简洁了，表达也更连贯。

译文：The test specimens were all labeled in order to identify their main parameters.

此外，科技文本常常使用数字、符号、公式、列表、表格和图形等副文本帮助阐释某个概念、原理或过程等。这些副文本使科技文本的语言表达具体、形象而又精炼，富有表现力。这些副文本体现了文本的专业性，清楚地阐释了抽象晦涩的概念和原理。若没有副文本，科技文本的表达力将会大大减弱，因此，丰富的副文本是科技文本的一大特征。在翻译带有数字的句子时，可以使用括号对句子进行简化。

◇ 例 13

原文：改进后的模型准确率是 93.2%，高于 VGG-16 网络模型的 90.5%.

原译：The accuracy of the improved model is 93.2%, higher than that of the VGG-16 network model, which is 90.5%.

解析：原译中有一个定语从句，提供的是百分比数据 90.5%，而这个没有必要用句子体现出来，从句的使用使句子变得复杂冗长。所以，将从句提供的数据置于括号中即可。

译文：The accuracy of the improved model is 93.2%, higher than that of the VGG-16 network model (90.5%).

◇ 例 14

原文：然后将样品放入一个温度恒定和湿度恒定的盒子中。温度设置为30℃，相对湿度设置为85%。

原译：The sample was then placed into a temperature-constant and humidity-constant box. The temperature was set at 30℃, and the relative humidity was set at 85%.

解析：原文包含两个句子，讲的是温度、湿度恒定，第二部分提供了两个数据，是对第一部分温度和湿度的具体说明，在这样的情况下，没有必要用一个单独的句子来进行解释。译文将该数据置于相应的参数后面，也就是 temperature-constant (30℃)，humidity-constant (85%)。如此一来，句子的长度不仅大幅缩减，而且意思也非常清晰。

译文：The sample was then placed into a temperature-constant (30℃) and humidity-constant (85%) box.

因此，如果翻译涉及数据，如频率、温度、电压、电流等的科技文本，可以考虑使用括号来简化句子的表达。括号不仅适用于简化涉及数字的科技翻译，也适用于出现多个并列项的句子。

◇ 例 15

原文：从转角空间的空间尺度和空间形态两个方面对地下商业街转角空间的构成要素进行整理，得出其主要的构成要素。

原译：Sort out the components of the corner space of underground commercial streets from two aspects of spatial scale and spatial form to obtain its main components.

解析：原译中 two aspects of spatial scale and spatial form 使用的 of 短语过长，其后不定式表目的的意味被弱化。可使用括号代替 of 短语，不仅简化句子，还在保证流畅性的前提下使句子主干更加突出，简洁明了。

译文：Sort out the components of the corner space of underground commercial streets from two aspects (spatial scale and spatial form) to obtain its main components.

练习
practice

1. 简答题。

(1) 语言简洁是否单指句子长短？谈谈你的理解。

(2) 如何实现汉英科技翻译中的语言简洁？

2. 赏析下列翻译，指出译文的优点。

(1) 水电解时分解为氢和氧，氢出现在负极周围，而氧出现在正极周围。

In electrolysis the water breaks up into hydrogen and oxygen, the hydrogen appearing about the cathode and oxygen about the anode.

(2) 在无线电系统的工程设计中，必须把这些考虑进去，并估计到它们的影响。

They must be considered and allowances made for their effects when engineering a radio system.

(3) 其答案不是像数字计算机那样取决于计算，而只是让物理定律自身起作用。

The answer is determined not by calculating, as in a digital computer, but by simply letting the laws of physics run their course.

3. 翻译下列句子，注意语言的简洁性。

(1) 经济中信息的应用主要体现在通讯的网络化上，也就是所谓的网络经济。

(2) 例如，用电话传送器把声能变成电能就是一种调制形式。

(3) 破坏宇称守恒会导致所有系统中都产生一个电偶极矩。

第二节　表达清晰准确

虽然科技文本的语言多为朴实客观的陈述性语句，但科技翻译绝不仅仅是数据的简单堆砌，因为除了保证数据准确无误外，语言表达的准确性和清晰性也是科技翻译中值得重视的两大要素，要做到不使读者费解，更要做到不让读者误解。

科技翻译的主旨就是传递科技信息、传播科技知识、进行科技交流，不折不扣地报道事实，确保科技信息传递的真实性，带有鲜明的"信息型"文本功能特征。对此，诺德(2001)指出，"信息型文本的主要功能在于向读者提供真实世界的客观事物和现象，语言和文体形式的选择应服从于这一功能"。纽马克(2001)也认为"语言的信息功能核心是外部情景、某话题的事实、语言外的现实，包括报道的观点或理论"。因而，对于"信息型"文本的翻译，为确保向译文读者提供真实世界的客观事物和现象，功能翻译学派认为应以"交流信息""始终如一地传达文本信息"为第一要义，其首要的评判标准在于，必须确保内容和信息充分展示于目标语之中，同时，以目标语为导向的彻底性则是其第二标准(Reiss，2004)。要达到这两个标准，译文表达必须客观精准(material accuracy)、信息明确(precise information)和语言合时(current language)。也就是说，译文的形式要从根本上顺从于目标语的习惯用法，要为译文读者所熟悉，使读者感觉不出在读译文。可见，信息传递的效果、内容的精确与表达的规范是科技翻译的核心和基准，功能对等、信息准确真实是科技翻译的第一要义。要做到这一点，译文语言必须客观准确、明白易懂并具有可读性，使译文读者基本上能以原文读者理解原文的方式从译文中获取相关科技信息。从这个意义上来说，奈达的功能对等原则特别适合科技翻译。因为，功能对等意味着译文必须使用一种共同语言，即一种最切近的自然对等语言再现原文的信息和内容。体现在科技翻译中，就是一种客观精确、真实自然、符合科技规范的标准语言，任何违背科学、不合逻辑、生搬硬套、含糊晦涩的语言表达都是科技翻译之大敌。

科技翻译的语言并不需要过多精雕细琢的华丽辞藻，但是语言表达在科技翻译中的作用也不容小觑，因为它影响着读者的感受。不清晰的语言表达会使原文中某些微妙的信息和态度在译文中丢失，甚至削弱原文的说服力。这就要求译者在科技文本翻译中应当尽量避免使用模糊的表达方式，而应该使用简单明了的语言，以确保科技文本的基本属性——清晰。因为科技文本通常需要精准地传达信息，使用含糊不清的表达方式可能会导致读者产生误解或理解不清。因此，在翻译时应该注重使用简单、明晰、准确的语言，以确保原文信息准确无误、有效传达给目标语读者，提高科技文本的可读性和实用性。

为实现清晰准确的表达，需要译者在行文的"信"(或"达")与"雅"之间寻求平衡，当二者无法兼顾时，应当首先考虑译文的"信"与"达"，做到有效表达原文信息。可以通过以下几个例子来体会"清晰准确"的优先级以及如何把握这种平衡。

【经典赏析】

原文：以上分析表明，并不是碳纤维越多，报文丢失率就越低。

译文：The above analysis shows that more carbon fibers do not make better loss performance.

赏析：对于"越……就越……"这一中文句式，英语中惯常的表达法是 the more…the more…本句如果采用这一表达，就意味着要将否定词 not 放在 more carbon fibers 之前，读者阅读起来有些费力，无疑会降低阅读的流畅性，且不符合英语表达习惯，有中式英语之嫌。鉴于此，可将 carbon fibers 作为从句主语，并添加动词 make，not 用来否定动词 make，这样更符合英语的表达习惯，句子焦点也清晰起来，读者易于理解。

【案例解析】

◇ 例1

原文：关于混凝土的研究应与诸如良好性能、多功能以及环境保护等许多发展方向平行开展。

原译：The research on concrete should be parallel with many development directions such as high performance, multi-function and environmental protection.

解析：原译将"平行开展"直译为 be parallel with，看似没有错误，实则是中式英语表达，会给目标语读者造成困惑。"平行开展"实际上就是"同步进行、二者同时开展"的意思。因此，可以对原文稍作意译，译为 conducted together with 即可，这样一来就准确清晰地传达了原文的含义。

译文：The research on concrete should be conducted together with many development directions such as high performance, multi-function and environmental protection.

从例1可以得到一些启发：虽然我们强调宁可用朴素的语言也要确保科技

文本准确清晰的基本属性，但不可失之偏颇，拘泥于原文或过于意译都不可取，要具体问题具体分析。当译文需要根据原文进行调整稍作意译时，更要谨慎选择最能体现原文含义的表达，一切以实现准确清晰的表达为目标。

◇ 例2

原文：ANOVA 分析结果显示，老年人的健康状况存在显著性差异，具体表现为随着年龄的增长老年人的健康水平下降，城镇的老年人健康水平高于农村老年人。

原译：According to ANOVA analysis, there are significant differences in the health status of senior citizens. Specifically, their health gets worse with the increase of age, and those living in urban areas enjoy better health than those in rural areas.

解析：senior citizens 确实可以指"老年人"，且与平时使用较多的 older people、the aged 相比似乎更高级，但用在此处并不准确。《牛津词典》对 senior citizen 的解释为 An older person especially somebody who has retired from work. People often call somebody a senior citizen to avoid saying that they are old or using the expression "old age pensioner"。也就是说，senior citizens 尤指那些退休老人，而观察原文的语境，显然 ANOVA 分析的对象绝不单指已经退休的老年人，而是根据一个年龄标准界定的老年人群体，与是否退休无关。因此这里将老年人译为 senior citizens 并不准确，会使目标语读者产生误解，科技文本的准确性也就大大削弱了。只需简单译为 the elderly 即可。

译文：According to ANOVA analysis, there are significant differences in the health status of the elderly. Specifically, their health gets worse with the increase of age, and those living in urban areas enjoy better health than those in rural areas.

◇ 例3

原文：由于支持向量机(SVM)不能挖掘具有复杂数据结构的用户消费记录和上下文信息，Tang 提出了一种基于回归预测的算法，大大提高了电子商务用户忠诚度评价的准确性。

原译：In view of the fact that the Support Vector Machine (SVM) cannot mine the user consumption records and context information with complex data structure, Tang proposed an algorithm based on regression prediction, which greatly improves the accuracy of e-commerce user loyalty evaluation.

解析：原译使用了含同位语从句的介词短语 in view of the fact that，表面上很
高级，但是表达太复杂，使同位语从句在该句中引出原因的意味被弱化，
整个句子重点不突出。因此，应进行简化，使用一个简单的 as 既能清
晰明确地提醒读者该句的表"因"功能，使读者迅速锁定重点，还大大
简化了句子。

译文：As the Support Vector Machine (SVM) cannot mine the user consumption
records and context information with complex data structure, Tang proposed
an algorithm based on regression prediction, greatly improving the accuracy
of e-commerce user loyalty evaluation.

练习
practice

1. 简答题。
(1) 导致表达模糊的原因有哪些？
(2) 在科技翻译中如何做到表达清晰准确？
2. 赏析下列翻译，指出译文的优点。
(1) 根据这个定律，在恒定温度下，某定量气体的压力和该气体的体积成
反比。

According to this law, the pressure of a given mass of gas at constant
temperature is in inverse proportion to the volume occupied by the gas.

(2) 费用通常是基于工程总造价的一个百分比，大约在 5%～10%的范围，
视工程大小和复杂程度而定。

The fee is usually based upon a percentage of the total construction cost of
the project, ranging from about 5 percent to 10 percent at a function of
project size and complexity.

(3) 对噪声的分类方法有许多种。可以根据具体情况对噪声的类型、噪声
源、噪声的影响或与接收机的关系进行细分。为了方便起见，我们在
这里把噪声分成两大类：其声源处于接收机外部的噪声和在接收机内
部产生的噪声。

There are numerous ways of classifying noise. It may be subdivided
according to type, source, effect, or relation to the receiver, depending on
circumstances. It is most convenient here to divide noise into two broad

groups: noise whose sources are external to the receiver, and noise created within the receiver itself.

3. 翻译下列句子，注意使用清晰准确的表达。

(1) 然而，由于数据更新的间隔很短，在如此短的时间间隔内错误的输出值对过程操作造成严重影响的可能性降低，因此，风险被最小化。

(2) 根据表 1，在过去五年里人们的食品结构发生了显著的变化。谷类食品的消耗量呈递减态势，而牛奶、水果、蔬菜类的健康食品消耗量却稳步增长。

(3) 据报道，此项目有 70% 的利润。

第三节　统一性原则

句子的统一性原则要求一个句子围绕一个主题展开，并表达一层基本意思。如果一个句子的所有组成部分都清晰地表达了同一个主题，就可以认为这个句子遵循了统一性原则。但这并不是说所有的句子都必须处理成简短的简单句。当几个句子表达同一个核心含义时，为了避免语意重复，就需要将有联系的句子联合，增强句子的完整性，使之表意准确，表达简洁。而当一个句子表达多个含义时，则需要将句子拆分，使表达的信息更为清晰，使句子清楚易懂。

【经典赏析】

原文：在 X 光下，金属顶端清晰可见，人们可以借助 X 光精确地观察和引导金属顶端进入心脏的全过程。

译文：The metal tip is very clearly visible under X-rays, so its progress into the heart can be watched and guided very precisely.

赏析：原文的两个部分之间存在因果关系，因为金属顶端在 X 光下清晰可见，所以人们才能借助 X 光观察金属顶端进入心脏的全过程。所以，不可将两部分译为两个独立的句子，可以在中间加逻辑连接词 so，这样就把两个句子合并在一起，表达更有力。

【案例解析】

◇ 例 1

原文：贝内特证明了在患有肌营养不良症的大白鼠的细胞里，钙含量调节机制是异常的。该发现有助于证实此疾病病变的关键可能在于神经系统这一存在已久的猜测。

原译：Bennet has shown that the calcium-regulating mechanisms are abnormal in the cells of mice suffering from muscular dystrophy. This discovery helps to confirm the long-held suspicion that this disease may be centered in the nervous system.

解析：原文中的"该发现"指前一句所说的"贝特证明了患病大白鼠细胞里钙含量调节机制是异常的"，因此，这两个句子表达的是同一个核心含义，且不难发现，二者之间还存在一定的因果关系，或者说第二句是对第一句内容的总结，所以可以将两个句子合并。

译文：Bennet has shown that the calcium-regulating mechanisms are abnormal in the cells of mice suffering from muscular dystrophy, thus helping to confirm the long-held suspicion that this disease may be centered in the nervous system.

◇ 例2

原文：对于中国的人工智能大模型，自然语言处理仍然是最活跃的研发领域，多模态次之。

原译：For China's large-scale AI models, natural language processing is still the most active field of research and development, and the multimodal field ranks second.

解析：可以看出，原文中第二句是对前一句内容的补充。原译把有联系的两个句子割裂，没有形成一个完整的统一体，对比之下，符合统一性原则的译文更加顺畅。

译文：For China's large-scale AI models, natural language processing is still the most active field of research and development, followed by the multimodal field.

◇ 例3

原文：干细胞是所有这些维持生命物质的起源。它是一种散布在骨髓中的血质胚芽，只要接到指令，便能发育成这三种不同类型的血细胞中的一种。

原译：The stem cell is the origin of all these life-sustaining agents. It is a sort of hematological embryo that idles in the marrow until ordered to develop into one of the three discrete blood cell types.

解析："它"就是指前一句的主语"干细胞"，也就是说，原文的两个句子主语相同，此时我们可以省略 be 动词以及第二个句子的主语，用逗号将两个句子合并，使句与句之间更加连贯紧凑，增强句子的完整性。

译文：Each of these life-sustaining agents begins as a stem cell, a sort of hematological embryo that idles in the marrow until ordered to develop into one of the three discrete blood cell types.

值得注意的是，在合并句子的时候要考虑两个信息之间的逻辑关系，这样才能使所合并的句子符合逻辑。同时，在合并成复杂句时要注意各种从句句式的应用，在合并成简单句时也需要注意特殊短语的使用。

根据如何合并的逆向思维，将句子拆分，也就是将句子拆分为几个简单句。句子拆分的方法较为简单，首先要将整个句子里面的核心含义分开，明确要拆分成几个句子，其次要注意拆分后的主语指代的是什么，明确这两点后，就可以进行拆分了。当一个句子表达多个含义时，将句子拆分可以使表达的信息更清晰，使句子更加清楚易懂。

◇ 例4

原文：保守治疗时间长，创面易感染加深，愈合后瘢痕严重，造成外观及正常生理功能损害。

原译：The conservative treatment is time-consuming, the wound is likely to be infected and severe scars can be formed after healing, resulting in unattractive appearance and damage to normal physiological function.

解析：原译按照中文断句将四个句子通过逗号、连词 and 以及一个非谓语结构译为一句。实际上可以对该句进行合理拆分，将原译的长句拆分成两个独立的句子，并根据上下文语境及句间逻辑适当添加过渡词(如表示原因、递进、并列等关系的副词)，避免冗长和复杂的句子结构，从而更好地传达信息和思想，使读者更加容易理解文本的内容，提高译文的质量和可读性。

译文：The conservative treatment is time-consuming and the wound is likely to be infected. In addition, severe scars can be formed after healing, resulting in unattractive appearance and damage to normal physiological function.

◇ 例5

原文：本研究将泡沫混凝土引入传统的吸波建筑材料中，并通过实验研究了炭黑和碳纤维作为吸波剂对泡沫混凝土基体结构稳定性和吸波性能的影响。

原译：In this study, foam concrete was introduced into traditional wave-absorbing building materials, and experiments were carried out to investigate the influence of Carbon Black (CB) and carbon fiber as wave-absorbing agents on the structural stability and wave-absorbing performance of the foam concrete matrix.

解析：虽然原文在两句之间使用了连词"并"，但英译时不必把两句用 and 相连，因为原文第一句"本研究将泡沫混凝土引入传统的吸波建筑材料中"结束时，意思已经相对完整，其核心含义为"泡沫混凝土的引入"，且第二句"并通过实验研究了……"与第一句并无逻辑关联，因此，可以直接在这里将句子拆分。如此一来，不仅保证了语义清晰，句式长短富有变化，还给目标语读者留出稍作思考的时间。

译文：In this study, foam concrete was introduced into traditional wave-absorbing building materials. Experiments were carried out to investigate the influence of Carbon Black (CB) and carbon fiber as wave-absorbing agents on the structural stability and wave-absorbing performance of the foam concrete matrix.

练习 practice

1. 简答题。

(1) 如何理解统一性原则？

(2) 如何判断句子需要做合并还是拆分？

2. 指出并改正下列句子中的错误，优化句子使其更具统一性。

(1) Every subject has a defining benefit to us students, which is exemplified by math can improve our logical thinking and art can develop our cognitive and creative skills.

(2) People tend to live in a nuclear family. So, it can cause problems for the elderly.

(3) Highways and subways have been constructed. They make it possible for people to travel from one place to another.

3. 判断下列句子是否遵循了统一性原则，如果不是，请对其进行修改。

(1) The silence of the forest was more oppressive than the heat, and at this hour of the day there was no sound at all, not even the whine of insects.

(2) People with seasonal affective disorder can be treated by using light boxes.

People are exposed to high doses of artificial light to mimic the brighter spring and summer mornings in the box.

(3) It can be concluded that the Baud rate is very important to the telephone engineer. Because this rate establishes the type of telecommunication channel to be used.

第四节　长短适中

英语中所说的长句一般指的是各种复杂句，其中还可能有多个从句，从句与从句之间的关系也是错综复杂的。简单来讲，英语中长句的形成在于各种成分的嵌套以及特殊形式结构的使用。

一般来说，英语长句具有如下几个特点：

(1) 结构复杂，逻辑层次多；

(2) 常须根据上下文作词义的引申；

(3) 常须根据上下文对指代词的指代关系做出判断；

(4) 并列成分多；

(5) 修饰语多，特别是后置定语很长。

科技性文体准确、规范，对思维的严谨性和逻辑性要求很高，长句完美地适应了这个要求，因而出现频率很高。在科技文体中表达复杂概念，特别是当这些概念涉及因果、对比、时间等逻辑关系时，使用长句可以补充更多的信息。但是，科技翻译中并不是使用长句越多就越好，原因有二：一是英语表达多富于变换，不管是在用词、语态、结构还是句式长短上都应当避免重复，使用多种多样的表达，长句和短句灵活运用；二是英语的长句中修饰语、并列成分、语言结构层次繁多，非常考验使用者的语言功底和逻辑思维，稍有不慎就会出现错误，同时也有可能不利于读者的理解。

也就是说，长句的使用并不一定比短句好。如果一个句子可以用短句表达清楚，那么就没有必要刻意使用长句。

【经典赏析】

原文：数据表明，外宣翻译的研究仍然是一个重点领域，其与国家政策和国际形势密切相关，学者们一直在关注该领域的研究，并取得了富有成效的研究成果。

译文：The data shows that studies on publicity translation remain a focused area, and are closely linked to the state's policy and international situation. Scholars have been continually paying attention to publicity translation research, with fruitful results obtained.

赏析：本句中包含的信息点较多，根据中文的表达习惯句子使用了多个逗号，在翻译为英文时，如果沿用这一习惯，译为一个英文句子，势必导致句子过长，表意也不清晰。因此，要对原文进行分解，合理地断句，使之表意清晰且符合英语的表达习惯。

【案例解析】

◇ 例1

原文：问卷中的每一组描述根据评估量表分为五个层次(非常差、差、不清楚、整体良好、非常好)，实验对象需要根据自己的心理认知从其中选择合适的层次。

原译：Each group of descriptors in the questionnaire is divided into five levels according to the evaluation scale (very poor, generally poor, ambiguous, generally good, very good), and the subjects are required to choose the appropriate level from the five levels in accordance with their psychological perception.

解析：原译中两个分句的修饰语都很长，第一个分句的括号里又是一长串对 scale 的解释说明，两个长句子拼凑在一起显得呆板沉闷。两个分句虽然都提到了"评估量表的 5 个层次"，但二者之间没有明显的逻辑关联，这种情况下，没有必要连续使用两个长句，完全可以将它们断开，形成两个独立的句子。如此一来，既保证了语义的完整，还使句子重点突出，句式长短变换有致。

译文：Each group of descriptors in the questionnaire is divided into five levels according to the evaluation scale (very poor, generally poor, ambiguous, generally good, very good). The subjects are required to choose the appropriate level from the five levels in accordance with their psychological perception.

◇ 例2

原文：中国海油集团 6 月 1 日宣布，我国海上首个百万吨级二氧化碳封存工程在南海东部海域正式投用，累计封存量将超 150 万吨，相当于植树近

1400 万棵。

原译：The China National Offshore Oil Corporation (CNOOC) announced on June 1 that China's first offshore million-tonne carbon storage project was put into operation in the eastern part of South China Sea and that it will store a total of more than 1.5 million tonnes of carbon dioxide (CO_2), which is equivalent to planting nearly 14 million trees.

解析：原译动词后有两个并列的宾语从句，句子形式单一，显得冗长乏味。应考虑在语义相对完整的地方以句号断开，增加文章的可读性。

译文：China's first offshore million-tonne carbon storage project was put into operation on June 1 in the eastern part of South China Sea, according to the China National Offshore Oil Corporation (CNOOC). The project was designed to store a total of more than 1.5 million tonnes of carbon dioxide (CO_2), which is equivalent to planting nearly 14 million trees.

◇ 例 3

原文：仅添加 1.0wt.% 的炭黑就可以将电阻率降低到 804 Ω·m，当进一步增加到 2.5wt.% 时，电阻率降低到 242 Ω·m，此时电导率为 0.00412 s/m。

原译：A mere 1.0wt.% addition of carbon black can reduce the resistivity to 804 Ω·m; and when the addition is further increased to 2.5wt.%, the resistivity decreases to 242 Ω·m, and the electrical conductivity is 0.00412 s/m at this point.

解析：原译使用分号连接两个句子。英语中分号的使用规则较为复杂，关于"分号后面是否能使用 and"这一问题尚存在争议。通常认为分号后不再需要任何连接词(如 and、but、or、nor 等)，但也有人认为只要分号使句子结构更清晰，分号后也可以使用 and，且在美国大学入学考试 SAT(Scholastic Assessment Test)和 ACT(American College Testing)中，阅卷者也并不在意这些细节，因为他们认为"This is a matter of style, not grammar."此外，尽管分号后使用 and 的例子也很多见，但近年来也有一些英语国家的学者认为，写作中使用分号已成为过时的用法。所以分号目前并没有统一认可的使用规则，为保证科技文本的客观性、真实性，应避免使用这类规则模糊的标点符号以及有争议的用词、语法等，以免削弱文章的说服力。此处与其使用分号表示停顿，不如直接使用句号断开，也能达到使句子结构清晰，句式长短变化多样的目的。

译文：A mere 1.0wt.% addition of carbon black can reduce the resistivity to 804

$\Omega \cdot m$. When the addition is further increased to 2.5wt.%, the resistivity decreases to 242 $\Omega \cdot m$, and the electrical conductivity is 0.00412 s/m at this point.

可以发现，以上三个例子都因为拘泥于原文的断句而不自觉地使用了长句，原文用逗号，译文也一直用逗号，也就是常说的"一逗到底"，这是翻译中容易产生的一种惯性思维。忠实于原文固然重要，但过度忠实反而影响了译文的表达效果。这三个例子也启发我们对翻译策略或规则的使用应当适度，要有自己的理解与把握，有时打破一般规则或许能带来不一样的效果，应当"通晓规则而不受限于规则"。

练习 practice

1. 简答题。

(1) 科技翻译中长句越多越好吗？谈谈你的理解。

(2) 长句和短句各有何优点？

2. 赏析下列翻译，指出译文的优点。

(1) 飞机靠机载系统操作，由一名地面观察飞行员进行控制。

The flights are operated by on-board systems supervised by a ground-based observer pilot.

(2) 放射状单元能让护士看到病人，因此能用更多的时间对他们进行护理。

Radial units allow nurses to visually supervise patients and spend more time on patient care.

(3) 与最初的 Galaxy S 一样，这些后续推出的产品都是高端机，尺寸时尚，运行速度快。

Like the original Galaxy S, these successors are high-end devices with stylish dimensions and fast chips.

3. 翻译下列句子，注意灵活运用长短句式。

(1) 该国正在抵御通货膨胀，这是一个导致社会动荡和政局紧张的问题。

(2) 直到 2018 年夏季，政府的官方数据才开始承认通货膨胀的规模。

(3) 现在绝大多数美国高校，不论是公立还是私立，都禁止携带枪支。

第十章　科技翻译篇章处理

　　篇章是结构和翻译的最大单位，译者的目标就是求得译文以句子和语段为基础的篇章与原文意义相符、功能相当。语言学家对篇章有不同的定义。传统汉语语法学家认为篇章就是"结词成句、集句成段、连段成章"，篇章是表达一个完整思想的单位，可以交代一件事情或事情的一部分，可以描写一个人物或景物，也可以针对某一问题发表看法等。韩礼德 (M. A. K. Haliday) (1976) 认为"篇章是一个语义单位"。美国学者罗伯特•博格兰德 (Robert de Beaugrande) (1981) 提出"篇章可以定义为符合衔接、连贯、有目的、可接受、含信息、含情景和互文性这七条篇章性标准的交际行为"。

　　在科技翻译中，要建立篇章翻译的宏观意识，即关注句子与句子、语段与语段、段落与段落之间的衔接与连贯，关注段落布局并在需要的情况下灵活进行局部调整，关注篇章用词造句的风格，关注语篇整体逻辑性等。此外，还要有篇章翻译的美学意识，充分认识到科技文本并非都是语言刻板、表述严肃、缺乏文采的，恰如其分的润饰语言，关注词法、句法及篇章的审美意识，可以增加文本的可读性。最后，还要关注篇章翻译的程式化，了解不同科技文本的词汇特征、句法特征、程式化结构等，做到规范化、专业化的翻译。

第一节　篇章翻译的宏观意识

　　科技文体描述客观世界及其规律，各种论点、论据以及论证互相依存，各部分通过内在的逻辑关系紧密融合，构成有机联系的语义整体。在翻译时，译者必须要关注文本词句的衔接和各部分的连贯，关注文本整体架构的统一。篇章翻译的宏观意识包含衔接、连贯、整体架构等，这些都是其逻辑性的体现。具体来说，篇章的衔接连贯依赖语段与语段或段落与段落之间的关系。衔接 (cohesion) 指的是借助词汇或语法手段使文脉相通，形成篇章的有形网络，常见的衔接手段包括同义词重述、语意贯通、照应、逻辑连接词的使用等。连贯 (coherence) 是指以信息发出者和接收者双方共同了解的情境为基础，通过逻辑推理来达到语义的连贯，形成篇章的无形网络。整体架构是指建立在逻辑关系基础之上、由逻辑结构显现的所有语义关系，译者只有准确理解语义关系的整体架构，才能对译文宏观把控，精准传达语义。

科技翻译中，译者首先要明确，任何词、句、段落只有和上下文相互衔接，形成连贯的语篇，才具有确定的意义。译者对原文局部或整体的理解也不能停留在单一的词或句子的层面上，而要从语篇层面来宏观考察词、句、段在整个语篇中的作用和意义，运用概念、判断、推理等手段来挖掘事物内部的本质联系及规律，做到概念明确、判断恰当、推理正确，构建出最接近原文的译文逻辑框架，确保译文脉络清晰、主次有序、层次分明、与原文等效。

【经典赏析】

原文：

第一章我们讲解了 PSTN 如何在电话网络用户之间建立通话。然而，事情并没有这么简单。首先，通常各国都有数个 PSTN，分属于不同运营商，只不过有些可能覆盖面积有限或只能为特定客户(通常为企业)服务。此外，很多国家的有线电视运营商除了播放电视节目，还能提供通话服务。当然，几乎所有国家都有不止一个移动电话网络，与固定 PSTN 有所区别。这些网络都需要互相连接，以便用户呼叫其他网络的用户。因此，本章简要介绍各类电话网络及其互联方式。

尽管固定 PSTN 和移动网络最初是用于语音通话，它们也能让用户把电脑接入互联网。因此，为了解释电话网络互联，本章也简要介绍互联网的概念以及电话网络如何提供联网服务。

本章要介绍的第二点，是对专用网络的需求。有了专用网络，国家电话运营商除了提供简单通话服务之外，也可以提供多种多样的其他服务。本章介绍这类专用网络，同时用一个简单模型帮助读者理解它们相互连接、相互支持的方式。

综上所述，显然各国的多个网络都以某种适合的方法相互连接，并能够向全世界的用户提供多种通话连接方式，接下来我们将介绍这一复杂情境。

译文：

In Chapter 1 we considered how a PSTN establishes telephone call connections between subscribers on its network. However, this is only part of the story. First, there are usually several PSTNs in any one country, each owned by different operators, although some may have a limited geographical coverage or serve only certain (typically corporate business) customers. Many countries also have Cable TV operators who provide telephone service, in addition to broadcasting TV distribution. Then, of course, nearly all countries now have more than one mobile telephone network, which are separate from the fixed PSTNs. All these networks need to interconnect so that their subscribers can call subscribers on any of the other networks. This chapter, therefore, briefly describes the various telephone networks and how they all interconnect.

Although the fixed PSTN and mobile networks are primarily designed for voice communications, they do also enable subscribers to connect their computers to the Internet. Thus, to complete the interconnect story, this chapter briefly introduces the concept of the Internet and how access to it is provided by the telephone networks.

The second aspect of the story is the need for specialist networks which enable a typical national telephone operator to provide a wide range of services beyond simple telephone calls. This chapter briefly reviews these specialized networks and uses a simple model to help explain how they link and support each other.

So, clearly there are many networks in each country all linking appropriately in order to provide a variety of types of call connections between subscribers across the world. We will now try to understand this complex picture.

赏析：译文中 this is only part of the story、to complete the interconnect story、the second aspect of the story 分别出现在文章的前、中、后三个位置，相互呼应，共同谱写末尾的 this complex picture。逻辑上，第一段的 several、also、more than one 相互呼应，说明了在电话网络用户之间建立通话在实际操作时的复杂性。全文中，although、therefore、thus、so 等连接词或副词表意清晰，逻辑严谨，各部分之间衔接紧密，构建了清晰的篇章网络，更符合英语的形合模式。

【案例解析】

◇ 例 1

原文：图 1.1(a)展示的是一个简单的双人单向电话电路，其结构包括说话者一端的麦克风，它与听话者一端的接收器通过电路相连，由电池为麦克风及接收器供能。对话期间，说话者声道振动造成了空气气压变化，即声波。声波从说话者处到达麦克风，被麦克风转换为与声波模态一致的电信号。

原译：Fig. 1.1(a) illustrates a basic simple one-way telephone circuit between two people, whose set-up comprises a microphone associated with the speaker, which is connected via an electrical circuit with a receiver at the remote end associated with the listener and a battery provides power for the operation of the microphone and receiver. During talking, variations in air pressure are generated by the vocal tract of the speaker, which are sound waves. Sound waves travel from the speaker to the microphone and are converted into an electrical signal varying in sympathy with the pattern of the sound waves.

解析：原译第一句话完全按照中文结构进行翻译，连用两个定语从句和一个并列句进行连接，逻辑层次不明晰，结构烦琐。译文把其分为单独的四句话，结构更为清晰。原译第二句话将"即声波"放在句子最后，用 which are sound waves，指代模糊，译文将其转移到下一句话，将 variations 用作主语，"即声波"用 known as sound waves 表示，语意连贯清晰。原译最后一句中 travel 和 are converted 并列，一为主动一为被动，语意跳跃，不利于理解。译文采用将 travel 作为主动词，converts 转为从句动词的方式，层次更加分明。上述例子在翻译时宏观考虑语篇的整体结构和表意，不拘泥于原文，灵活处理，译文也更加连贯通顺。

译文：Fig. 1.1(a) illustrates a basic simple one-way telephone circuit between two people. The set-up comprises a microphone associated with the speaker, which is connected via an electrical circuit with a receiver at the remote end associated with the listener. A battery provides power for the operation of the microphone and receiver. During talking, variations in air pressure are generated by the vocal tract of the speaker. These variations in air pressure, known as sound waves, travel from the speaker to the microphone, which converts them into an electrical signal varying in sympathy with the pattern of the sound waves.

◇ 例 2

原文：传输系统用于在两个相距遥远的节点之间提供链路，可以是单向链路，如无线电广播的信号只从发射机传输到无线电接收机；也可以是双向链路，如在通话连接中进行的是双向对话。如第一章所述，在电信网络中，节点之间的链路是双线或四线电路。在四线电路中，单独的去路和回路均被视作传输通道，可参见第三章图 3.14 中展示的多路复用传输系统的广义图。

原译：The purpose of a transmission system is to provide a link between two distant points or nodes, it can be unidirectional link, for example, in radio broadcast transmission is only from the transmitter to the radio receivers; it can also be bidirectional, for example, in a telephone connection it is a two-way conversation. As shown in Chapter 1, in telecommunication networks the links between nodes are either 2- or 4-wire circuits. In 4-wire circuits, the separate "Go" and "Return" paths are considered as transmission channels, and we can refer to the generalized view of a multiplexed transmission system in Fig. 3.14 of Chapter 3.

解析：原译完全按照原文句段进行翻译，第一句中使用了三个并列句，而译文则将其分译为两个句子，句子表意重心更加均衡，使用重复 link 一词的方法进行衔接，并在第二句中突出对比 unidirectional 和 bidirectional，将"例如"后面的部分转为 as in the case of 引出的介词短语，更加简洁，且有语言的对称美。原译中将 as shown in Chapter 1 按照中文句意放于句首，主要信息不突出，译文则将其放于句尾，重点突出。原译将最后一句译为两个并列句，且分别使用非人称主语和人称主语，从被动态突然转换到主动态，不甚连贯，译文将第二个并列句转为 as 短语结构，句子重要信息、次重要信息一目了然，且句意更为连贯。上述例子在翻译时灵活调整语篇的语句划分，不拘泥于原文，灵活处理，表意重点突出，译文更加连贯通顺。

译文：The purpose of a transmission system is to provide a link between two distant points or nodes. The link may be unidirectional, as in the case of radio broadcast where transmission is only from the transmitter to the radio receivers, or bidirectional, as in the case of a telephone connection with its two-way conversation. In telecommunication networks the links between nodes are either 2- or 4-wire circuits, as described in Chapter 1. With 4-wire circuits, the separate "Go" and "Return" paths are considered as transmission channels, as explained in the generalized view of a multiplexed transmission system in Fig. 3.14 of Chapter 3.

第二节 篇章翻译的审美意识

　　人们一般认为文学翻译重在激发读者情感，因此多用形象思维方式，语言更为丰富生动、形象具体，遣词造句旨在达到调动情绪的目的，而科技翻译旨在启迪读者的理性，要求客观准确地反映事物，因而要求使用的语言明晰简练，少用带有文学色彩的修饰语或修辞格。然而，科学和艺术不分家，科技翻译不但要有科技价值，也要有艺术的审美价值。

　　美并没有绝对的、永恒的标准。科技翻译中的美不是奔涌的文思、华丽的辞藻、跌宕起伏的情节、幽默诙谐的语言，而是体现在准确使用科技术语、流畅熟练地运用文体程式、缜密合理地衔接篇章、恰当运用修辞手段，充分展现科技文体的精确、整齐、简约、有序、和谐之美。精确是指科技文体在用词方面要求准确，尽量避免含混不清和一词多义，精准使用术语既可以增加内容的精确性，又能提升形式的简洁程度。整齐是指科技文体的客观性、准确性和严密性，需要采用合理的句式结构，如并列结构、从属结构等，且讲究对称之美。简约在科技文体的名词化结构中可见一斑，既可以避开人称主语以体现客观性，也可以避免句式结构臃肿。有序是指科技文体的规律性，特定的科技语篇的信息分布有一定的规律可循，呈现出一种有序美，如科技文体的程式化、主题句的使用、主述位推进等。和谐指的是科技文章逻辑严密，恰当使用语段衔接手段，即逻辑纽带、照应关系以及词汇纽带等，使语篇宏观结构具有和谐之美，有助于译者正确理解原文，使译文逻辑清楚、条理分明、浑然一体。

【经典赏析】

原文：如今，人们广泛认为电信是用电气的方式进行的远距离通信。电信最早的形式是 1837 年由惠斯通和莫尔斯两人各自独立发明的电报。电报是点对点的单向通信，要依赖受过训练的话务员在话语或文字与电报电缆发送的特殊信号之间进行互译。虽然有其缺陷，但是电报确实极大地推动了城市间铁路运行、新闻传播及私人信息的传递。电信一方面有实用性，另一方面又受限于需要训练有素的话务员，因此人们希望能有一种简单易用的双向语音通信技术，以便人人都可以使用。

译文：Telecommunications is today widely understood to mean the electrical means of communicating over a distance. The first form of telecommunications was that of the Telegraph, which was invented quite independently in 1837 by two scientists, Wheatstone and Morse. Telegraphy was on a point-to-point unidirectional basis and relied on trained operators to interpret between the spoken or written word and the special signals sent over the telegraph wire. However, the use of telegraphy did greatly enhance the operations of railways and the dissemination of news and personal messages between towns. This usefulness of telecommunications on the one hand and the limitation of needing trained operators on the other led to the aspiration for a simple means of bi-directional voice telecommunications that anyone could use.

赏析：译文将"虽然有其缺陷"简单译为 however，与前文的"要依赖"与后文的"受限于"照应，翻译更具整体性。最后一句将"有实用性"与"受限于"采用名词化变为 usefulness 和 limitation 作主语，省去原来的主语"人们"，加上 led to 作为谓语，其谓语动词"希望"也从动词变为名词，表示目的的"以便人人都可以使用"转为定语从句直接放在句尾，整个句式结构整齐、简约，呈现出有序的对称美。

【案例解析】

◇ **例 1**

原文：集中器交换机将线路占用率提高为原来的 10 倍，但这样可能导致用户打不通电话，因为同一时刻拨打电话的 2000 个用户中，只有 200 个用户的通话可以接通。当任何通话无法接通时，交换机就处于"拥塞"状态，无法接通的呼叫被视为"呼损"。显然，我们可以算出呼损的概率。一般来说，呼损概率较低，在一天中最繁忙的时间通常也只有 1%～2%。网络运营商需要在尽量保持可接受范围内的低呼损率和尽量降低设备及网络容量的成本之间寻求经济平衡，因此集中率会尽可能保持高位。

原译：Although the concentrator switch has improved the occupancy of the lines by a factor of 10, it may lead to the result that subscribers cannot get through because calls from only 200 of the 2,000 subscribers can be carried at any one time. When calls are unable to get through, the switch is deemed to be in "congestion" and these calls are considered as "lost". Clearly, we can calculate the probability of lost calls. Normally, this probability is low, typically only about 1 to 2 percent during the busiest time of day. The network operator has to economically balance between keeping the probability of lost calls as low as possible in an acceptable range and gaining as much cost saving in equipment and network capacity, and therefore the concentration ratio will be as high as possible.

解析：译文开头采用了 introduced the possibility 的表述，避开从句叠用带来的复杂句式，同时将"可能"一词巧妙嵌入其中，用 since 替换 because 表示上文提到的显而易见的原因，与 a factor of 10 照应，any calls 更忠实于原文的"任何通话"，接着用 the probability may be determined statistically 的被动句替换原本的 we can calculate the probability，避免中途将主语更换为人称主语，与前后文更为连贯，economic trade-off 比 economically balance 用词更为规范，in an acceptable range 直接转为 acceptable probability，更加简洁，最后的并列句转为 by 引导的方式状语，句子结构更加紧凑。整体来说，译文表意精准、结构严密、前后照应，读起来更觉语义顺畅。

译文：Although the concentrator switch has improved the occupancy of the lines by a factor of 10, it has introduced the possibility of subscribers not being able to get through since calls from only 200 of the 2,000 subscribers can be carried at any one time. For the time that any calls are unable to get through, the switch is deemed to be in "congestion" and these calls are considered as "lost". Clearly, the probability of lost calls may be determined statistically. Normally, this probability is low, typically only about 1 to 2 percent during the busiest time of day. The network operator has to make an economic trade-off between keeping the acceptable probability of lost calls as low as possible against gaining as much cost saving in equipment and network capacity by having as high a concentration ratio as possible.

◇ 例2

原文：虽然如此，但直到 20 世纪 70 年代末 80 年代初，我们才采用数字交换取代了模拟交换系统，先是用在长途交换机，然后扩展到本地交换机。引入数字交换的主要原因是能够全部升级为半导体电子交换，从而减少人工成本，降低投资成本，减少办公费用，提高连接清晰度，拓展新服务和新功能。模拟交换和传输系统已逐渐被数字系统所取代，自 20 世纪 90 年代中期以来，几乎所有发达国家的 PSTN 都是完全数字化的。

原译：However, until the late 1970s and early 1980s, digital switching was used to replace the analogue switching systems, first used in the trunk exchanges and later in the local exchanges. The primary reason for introducing digital switching was to be able to fully make semiconductor electronic exchanges, thus reducing manual operational costs, lower capital costs, reduce accommodation cost, improve clarity of the connections, and extend new services and features. The analogue switching and transmission systems have been replaced by digital systems, and since the mid-1990s nearly all PSTNs in the developed World are entirely digital.

解析：译文第一句直接采用 it was not until… that…结构，表意更为精准，initially… then later 表达更为精练，第一个 exchanges 也采用了省略结构。driver 比 reason 更能形象地表示变革的推动力，接着采用 consequential 表示"从而"的意思，使用"with + 名词化"的结构，名词短语依次并列，更加整齐划一，用 potential 和 clarity 并列，将原文的"提高"和"拓展"统一为 improvements。最后一句原译中采用并列结构的句子，而在译文中转为因果关系的主从句，结构更加紧密，语义更为连贯。译文整个句子选词精准，句式结构精妙，翻译灵活，既忠实于原文，又不失英语的美感。

译文：However, it was not until the late 1970s and early 1980s that digital switching was used to replace the analogue switching systems, initially the trunk then later the local exchanges. The primary driver for introducing digital switching was to be able to fully make semiconductor electronic exchanges, with the consequential reduction in manual operational costs, lower capital costs, less accommodation required, improvements in clarity of the connections and the potential for new services and features. Progressively, the analogue switching and transmission systems have been replaced by digital systems so that since the mid-1990s nearly all PSTNs in the developed World are entirely digital.

第三节　篇章翻译的程式化

文本的规范性和程式化是科技文体的重要特征，常规科技文献包含专利文献、科技期刊、学位论文、会议文献、技术档案、产品资料、科技图书、科技报告、政府出版物、标准文献等不同类型，同类的语篇具有大致相同的体例，如学位论文通常包括 Title、Abstract、Introduction、Methods、Result、Discussion、Acknowledgments、References 等部分，语言表达方式也呈现较为固定的模式。纽马克(1981)主张按照语言的功能将文本分为表达型、信息型和呼唤型三类。针对表达性文本，他主张以作者为中心，采取与原作风格相似甚至一致的语言进行翻译，以保留其"美学价值"，即采用语义翻译法；而针对信息型和呼唤型文本，他主张采用交际翻译法，即以读者为中心，使译文的语言尽量贴近读者的语言水平。科技文本主要属于后两种类型，其格式和写作套路相对固定，写作目的明确，译者应该准确把握原文的写作目的，了解文本的词汇特征、句法特征、程式化结构等，使译文具备相同的功能。

【经典赏析】

◇ 例1

原文：

产品售后保修条款

摩恩产品严格按照高标准的生产工艺制造，摩恩承诺为消费者购买的洁具产品提供有期限免费保修及终身维修服务。

龙头免费保修期为：阀芯 5 年，电磁阀感应器 2 年，其他附件如淋浴杆、软管、花洒、落水、把手、装饰盖、提拉杆、混水器等的保修期均为 1 年。均自消费者购买日(以摩恩产品分销商开具的发票或收据上载明的日期为准)起计。

免费保修范围为产品及产品配件(如淋浴杆、软管、花洒、落水、把手、装饰盖、提拉杆、混水器等)由于材质及工艺原因而造成的故障或损坏。免费保修期内，我们将无偿为您提供修理服务及更换发生故障的产品或产品配件。摩恩的售后服务责任仅限于对故障产品的修理，而不延及您因产品故障而发生的其他损失，法律另有规定的除外。

您需了解：产品外观碰伤、刻划痕迹，以及滥用及不正当保养所导致的故障或损坏，均不属于免费保修范围。即使在免费保修期内，属下列(1)～(5)情况之一的，需酌情收取修理费用：

(1) 不能出示相符的保修卡和有效购买凭证情况的；

(2) 未按使用说明书告知的使用方法和注意事项造成的故障；

(3) 自行拆卸、维修、改装所造成的故障和损坏；

(4) 由于不可抗力等外部原因引起的故障和损坏；

(5) 由于运输、搬卸、挤压等造成的损坏。

对于保修期外或不在保修范围内的产品，如需维修，摩恩将酌情收取上门费、人工费及材料费。

本保修条款仅对原始消费者和酒店行业的购买者有效，并不适于本产品在工业、商业、贸易和非酒店行业的使用。

如对本保修条款内容有任何疑问，请拨打 400-630-8866 咨询热线，或向当地分销商咨询。

安装指南(节选)

(1) 用生胶带缠上弯头(5)较小的一头，然后接上供水管(确保 3 圈以上螺纹连接)。要有足够的生胶带以密封弯头与水管间的连接处。调整两弯头位置水平，相距 150 mm，并凸出墙面 34.5 mm～39 mm。

(2) 将法兰罩(4)穿入弯头(5)，使法兰罩与墙壁平齐。将垫圈(3)放入阀体上的连接螺母(2)内，并用扳手将连接螺母与弯头拧紧。

译文：

MOEN LIMITED WARRANTY

We offer limited warranty and lifetime maintenance for Moen products, which are all manufactured with high standards of workmanship.

The period of warranty shall start from the date of purchase of the product (according to the invoice or receipt issued by Moen distributors) and shall cover a period of the following: the warranty period for cartridge is 5 years, the warranty period for solenoid valve sensor is 2 years, and the warranty period of other accessories such as shower bar, hose, shower, drain assembly, handle, cover, lift rod, aerator, etc. is 1 year.

This warranty is your coverage against material and workmanship malfunction or defects in Moen products or components (such as shower bar, hose, shower, drain assembly, handle, cover, lift rod, aerator, etc.). During the warranty period, we offer free maintenance and changing of malfunctioned products or product components. Moen's aftersales service liability is limited to the maintenance of faulty items and does not cover your other losses caused by product malfunction, except for cases otherwise stipulated by law.

Cosmetic damages (scratches, dents, etc.) and malfunction or defects caused by abuse or improper maintenance are not covered by this warranty. Although the warranty period remains valid, in case of the following conditions (1)–(5), the maintenance cost (material cost and labor cost) shall still be charged as the case may be:

(1)　Corresponding warranty card and valid invoice cannot be presented;

(2)　Faults were caused by the incompliance with the instructions and precautions provided in the user's manual;

(3)　Damages and faults were caused by self-actuated dismantling, repair and refit;

(4)　Faults and damages were caused by external factors such as force majeure;

(5)　Damages were caused by transportation, unloading and extrusion, etc.

This warranty is valid for the original consumers only (including hotel use), and excludes industrial and commercial use of the product (except for hotel use). This card is only responsible for product maintenance rather than any other liabilities.

In the event that the product requires maintenance but exceeds the warranty period, or that the product is not covered by this warranty, labor cost and material cost shall be charged as the case may be.

Please call 400-630-8866 or contact local dealer for any question.

INSTALLATION (excerpts)

(1)　Apply the teflon tape to smaller ends of inlet elbows (5) and thread inlet elbows into supply lines (3 pitches min thread connection). Use enough teflon tape to ensure a good seal between inlet elbows and supply lines. Rotate inlet elbows until they are level, 150mm apart, and protrude from the wall 34.5 mm–39mm.

(2)　Thread wall flanges (4) onto inlet elbows (5), tighten wall flanges until they touch the wall. Insert washers (3) into the connection nut (2) on the valve body. Thread connection nuts onto inlet elbows with a wrench.

赏析：

　　本语篇选自科技文体中的产品资料，第一部分为产品售后保修条款，第二部分为安装指南(节选)。第一部分的译文中多使用产品及产品部件相关专业词汇、法律相关词汇及句式表达等。第二部分译文的程式化体现在产品部件后使用带括号的数字进行图文对照说明，句式以祈使句为主，通常按照操作的先后顺序排列并带有数字序号，句子整体简洁明晰，未使用过于复杂的句式。

◇ 例2

原文：

(54) **可调节照明角度的扇形手电筒**

(57) 　　　　　　　　　**发明摘要**

可调照明角度的扇形手电筒包括设置在手柄前端的连接座、安装在连接座前端的前投影主灯以及分别旋转到主灯两侧的左前投影副灯和右前投影副灯。主灯和两个副灯并排，处于同一平面上，以扇形角度投射光线，两个副灯可以单独向外侧转动，以调整其照明角度和位置。

8 项权利要求，7 张图纸
可调节照明角度的扇形手电筒

发明说明

本发明属于一种手持式照明工具，更具体地说，这种手电筒的一个主灯和两个副灯并排处于同一平面上以近乎扇形的角度投射光线，两个副灯可以偏离主灯，以便调整照明角度和位置。

发明背景(正文略去)

发明概述(正文略去)

图纸的简要描述(正文略去)

优选的实施例描述(节选)

专利申请范围：

1. 一种具有可调节照明角度的扇形手电筒，具体部件包括：

手柄，供用户抓握，带有前端，手柄中安装有电池。

连接座，固定在手柄的前端，带有顶部，前端朝向前方，两个侧端分别位于连接座的两侧，连接座中安装有电路板，电路板与电池相连，连接座顶部安装有开关，开关与电路板相连。

主灯，安装在连接座的前端，由电路板供电，可将光线投射到前方。
两个副灯，分别安装到连接座的两个侧端，分别与主灯的两侧相对，由
电路板供电，分别向左前方和右前方投射光线；主灯和两个副灯并排，
处于同一平面上，以扇形角度投射光线，两个副灯可以在同一平面上扭
转，也可向主灯外侧打开，以调整其照明角度和位置；连接座带有底部，
底部有充电接口和多个充电指示灯，与电路板相连。

2. 根据权利要求 1 所述的可调节照明角度的扇形手电筒，其特征在于，
所述主灯带有与所述连接座相连的主灯灯壳，且所述主灯灯壳前端边缘
朝向前方，前端边缘装有主灯透光板，所述主灯灯壳中装有主灯反射器，
主灯 LED 与电路板相连，可借助主灯反射器和主灯透光板向前投射光
线；所述副灯带有与所述连接座相连的副灯灯壳，且所述副灯灯壳前端
边缘朝向左侧或右侧，前端边缘装有副灯透光板，所述副灯灯壳中装有
副灯反射器，副灯 LED 与电路板相连，两个副灯反射器和两个副灯 LED
分别面向左前方和右前方，与主灯 LED 呈一定夹角，并通过各自的副
灯反射器和副灯透光板分别向左前方和右前方投射光线。

译文：

(54) FAN-SHAPED FLASHLIGHT WITH ADJUSTING LIGHTING ANGLE

(57)   ABSTRACT

A fan-shaped flashlight with an adjustable lighting angle includes a
connection seat disposed at the front end of a handle, a front-projection
primary light installed at the front end of the connection seat, and two
left-front and right-front projection secondary lights pivoted to two sides of
the primary light respectively. The primary light and two secondary lights
are arranged side-by-side on the same plane to project light within a
fan-shaped angle, and the two secondary lights can be turned separately
towards the outside relative to the primary light to adjust their lighting angle
and position.

8 Claims, 7 Drawing Sheets
FAN-SHAPED FLASHLIGHT WITH ADJUSTING LIGHTING ANGLE
FIELD OF THE INVENTION

The present disclosure relates to a hand-held lighting tool, and more particularly to a flashlight with a primary light and two secondary lights which are arranged side by side on the same plane to project light within a substantially fan-shaped angle, and the two secondary lights can be reflected relative to the primary light in order to adjust the lighting angle and position.

BACKGROUND OF THE INVENTION
SUMMARY OF THE INVENTION
BRIEF DESCRIPTION OF THE DRAWING
DESCRIPTION OF THE PREFERRED EMBODIMENTS

What is claimed is:

1. A fan-shaped flashlight with adjustable lighting angle, comprising:

a handle, provided for a user's grip, and having a front end, and a battery installed in the handle;

a connection seat, fixed to the front end of the handle, and having a top side, a front end configured to be facing the front, and two side ends disposed on two sides of the connection seat respectively, and the connection seat comprising a circuit board installed therein and electrically coupled to the battery, and a switch installed at the top side of the connection seat and electrically coupled to the circuit board;

a primary light, installed at the front end of the connection seat, and supplied with electric power by the circuit board to project light towards the front;

two secondary lights pivoted to two side ends of the connection seat respectively, and configured to be opposite to two sides of the primary light respectively, and supplied with electric power by the circuit board to project light towards the left front and the right front respectively; wherein the primary light and the two secondary lights are arranged side by side on the same plane to project light within a fan-shaped angle, and the two secondary lights can be turned on the same plane and spread open towards the outside relative to the primary light to adjust the lighting angle and a position of the two secondary lights, and wherein the connection seat has a bottom side, with the bottom side having a charging interface and a plurality of charging indicator lights separately and electrically coupled to the circuit board.

2. The fan-shaped flashlight with an adjustable lighting angle according to claim 1, wherein the primary light comprises a primary lamp housing coupled to the connection seat, and the primary lamp housing has a front edge facing the front, and the front edge of the primary lamp housing comprises a primary light-transmitting plate, a primary reflector installed in the primary lamp housing, and a primary LED electrically coupled to the circuit board, and provided for projecting a light through the primary reflector and the primary light-transmitting plate towards the front; each secondary light comprises a secondary lamp housing pivoted to the connection seat, and the secondary lamp housing has a front edge facing the left front or the right front, and the front edge of the secondary lamp housing has a secondary light-transmitting plate, and the secondary lamp housing comprises a secondary reflector installed therein, and a secondary LED electrically coupled to the circuit board, and the two secondary reflectors and the two secondary LEDs configured to be facing the left front and the right front with an inclined angle relative to the primary LED, and emitting a light through the corresponding secondary reflector and secondary light-transmitting plate and projecting the light towards the left front and the right front.

赏析:

　　本语篇节选自科技文体中的专利文献,这里选取了 abstract、field of the invention 和 description of the preferred embodiments 中的部分段落。专利文献具有法律效力,语篇结构由固定的部分组成,缺一不可,每一部分均有较为固定的语言表述和句式,用词专业规范,如本发明中各部分的部件采用规范表述,如 lamp housing、circuit board、connection seat 等。此外,专利文献特有的一些表述,如 the present disclosure、wherein、therein、claim 等都有固定的译法,原文和译文均为正式的书面语体。

练习
practice

1. 简答题。
(1) 在对科技翻译篇章进行处理时,需要注意哪些问题?
(2) 在科技翻译篇章中,衔接和连贯分别指什么?

2. 赏析下列翻译，指出译文在篇章方面的优点。

(1) 天文观测有各种各样的形式，有时我们想要确定特定时间两个天体间的视距离，有时我们想知道一个天体在某一特定时间所处的位置，或是到达天空某一特定圆周的时刻，有时我们只想通过观察一个天体的表面测量它的亮度或者研究其光谱。因此，很多专门的仪器应运而生。

Astronomical observations are of various kinds: sometimes we desire to ascertain the apparent distance between two bodies at a given time, the position a body occupies at a given time, or the moment it arrives at a given circle of the sky, and sometimes we wish merely to examine the surface of a body to measure its light or to investigate its spectrum. For all these purposes, special instruments have been devised.

(2) 任何网络都是节点和链路组成的系统。我们的日常生活中有很多网络的例子：道路网络中的十字路口相当于节点，路口之间的道路就是网络的链路。同样，铁路网络中的铁轨(链路)连接了车站(节点)。航空网络中，机场是节点，飞机航线则是链路。

Any network is a system of nodes and links. There are many examples of networks in everyday life. We talk of the road network, where the junctions form the nodes and the stretches of road in between are the links of the network. Similarly, the rail network comprises the rail links joining the station nodes. A further example is that of the airlines network, where airports provide the nodal functions and the airline routes provide the links.

(3) 从一颗恒星发射到大物镜的所有光线都传送到了观测者的眼中(吸收和反射的损失忽略不计)，所以它接收的光量显然比自然获得的光量要多得多，其倍数差等于物镜与瞳孔面积相比的倍数差。因为光线的亮度与透镜直径的平方成正比，假设我们的瞳孔直径是 1/5 英寸，那么 1 英寸的望远镜会让光线增强为自然光的 25 倍，10 英寸的望远镜则会让光线增强为自然光的 2500 倍，而拥有 36 英寸大光圈的利克望远镜，光线会达到自然光的 32400 倍。然而，目镜的放大率会将光线扩散到一个更为宽广的角状区域，因此我们不能据此认为月球等呈现圆盘状的行星也按照上述比例增强光线亮度；事实上，我们可以证明任何光学装置都不能让一个展开面的光线亮度超过肉眼看到的光线亮度，但是，望远镜极大地增加了可利用的光总量，因此众多微弱到肉眼无法看到的恒星可以借助望远镜显现出来，尤为重要的是，用望远镜在白天更容易看到那些较亮的恒星。

Since all the rays from a star which fall upon the large object-glass are transmitted to the observer's eye (neglecting the losses by absorption and reflection), he obviously receives a quantity of light much greater than he would naturally get: its multiple difference is as great as the multiple difference between the area of the objective lens and that of the pupil of the eye. If we estimate this latter as having a diameter of one-fifth of an inch, then a 1-inch telescope would increase the light twenty-five times, a 10-inch instrument 2,500 times, and the great Lick telescope, of thirty-six inches' aperture, 32,400 times, the amount being proportional to the square of the diameter of the lens. It must not be supposed, however, that the apparent brightness of an object like the moon, or a planet which shows a disc, is increased in any such ratio, since the eye-piece spreads out the light to cover a vastly more extensive angular area, according to its magnifying power; in fact, it can be shown that no optical arrangement can show an extended surface brighter than it appears to the naked eye, but the total quantity of light utilized is greatly increased by the telescope, and in consequence, multitudes of stars, far too faint to be visible to the unassisted eye, are revealed; and, what is practically very important, the brighter stars are easily seen by day with the telescope.

3. 赏析节选自 2023 政府工作报告的中英双语版，总结译文的程式化特点。

过去一年和五年工作回顾
A Review of Our Work in 2022 and Over the Past Five Years

在攻坚克难中稳住了经济大盘，在复杂多变的环境中基本完成全年发展主要目标任务，我国经济展现出坚强韧性。

Overcoming great difficulties and challenges, we succeeded in maintaining overall stable economic performance. Amid a complex and fluid environment, we were able to generally accomplish the main targets and tasks for the year. Such achievements are a testament to the tremendous resilience of China's economy.

把年度主要预期目标作为一个有机整体来把握，加强区间调控、定向调控、相机调控、精准调控，既果断加大力度，又不搞"大水漫灌"、透支未来。

We adopted a holistic approach to achieving the projected development targets for each year and stepped up range-based, targeted, well-timed, and precision regulation. We took decisive measures to step up macro policy support while refraining from adopting a deluge of strong stimulus policies that would eat into our potential for future growth.

加大对易地搬迁集中安置区等重点区域支持力度，坚持并完善东西部协作、对口支援、定点帮扶等机制。

We stepped up support for areas where residents relocated from inhospitable areas were resettled together in communities as well as for other key areas, and continued to implement and improve mechanisms for east-west cooperation, paired assistance, and targeted support.

大道至简，政简易行。

Great truths are always simple, and simple government administration is always most effective.

严格执行环保、质量、安全等法规标准，淘汰落后产能。开展重点产业强链补链行动。

Laws, regulations, and standards on environmental protection, quality, and safety were strictly implemented, and outdated production facilities were shut down. We worked to shore up weak links in the industrial and supply chains of key industries.

围绕构建新发展格局，立足超大规模市场优势，坚持实施扩大内需战略，培育更多经济增长动力源。

To create a new pattern of development, we fully leveraged China's super-sized market and carried out the strategy of expanding domestic demand to foster more growth engines.

完善强农惠农政策，持续抓紧抓好农业生产，加快推进农业农村现代化。

We improved policies to boost agriculture and benefit farmers, continuously gave priority to agricultural production, and moved faster to modernize the agricultural sector and rural areas.

面对外部环境变化，实行更加积极主动的开放战略，以高水平开放更有力促改革促发展。

In response to changes in the external environment, we pursued a more proactive strategy of opening up and worked to boost reform and development with high-standard opening.

统筹能源安全稳定供应和绿色低碳发展，科学有序推进碳达峰碳中和。

We both ensured a stable and secure energy supply and promoted green and low-carbon development. We worked toward the targets of peak carbon emissions and carbon neutrality with well-conceived and systematic steps.

贯彻以人民为中心的发展思想，持续增加民生投入，着力保基本、兜底线、促公平。

Acting on a people-centered philosophy of development, we continued to increase inputs in areas important to people's wellbeing, meet people's basic needs, provide a cushion for those most in need, and promote social fairness.

坚持依法行政、大道为公，严格规范公正文明执法，政府的权力来自人民，有权不可任性，用权必受监督。

We exercised law-based government administration, served the common good of all, and saw that the law was enforced in a strict, procedure-based, impartial, and non-abusive way. The power of the government comes from the people; it should not be wielded as one pleases and the exercise of power must be subject to supervision.

对今年政府工作的建议

Recommendations for the Work of Government in 2023

把恢复和扩大消费摆在优先位置。多渠道增加城乡居民收入。

We should give priority to the recovery and expansion of consumption. The incomes of urban and rural residents should be boosted through multiple channels.

围绕制造业重点产业链，集中优质资源合力推进关键核心技术攻关。

We should, with a focus on key industrial chains in the manufacturing sector, pool quality resources and make concerted efforts to achieve breakthroughs in core technologies in key fields.

构建亲清政商关系，为各类所有制企业创造公平竞争、竞相发展的环境。

We should cultivate a cordial and clean relationship between government and business and create an environment in which enterprises under all forms of ownership can compete and grow on a level playing field.

开放的中国大市场，一定能为各国企业在华发展提供更多机遇。

With a vast and open market, China is sure to provide even greater business opportunities for foreign companies in China.

深化金融体制改革，完善金融监管，压实各方责任，防止形成区域性、系统性金融风险。

We need to deepen reform of the financial system, improve financial regulation, and see that all those involved assume their full responsibilities to guard against regional and systemic financial risks.

巩固拓展脱贫攻坚成果，坚决防止出现规模性返贫。

Our achievements in poverty alleviation should be consolidated and expanded to prevent large-scale relapse into poverty.

加强城乡环境基础设施建设，持续实施重要生态系统保护和修复重大工程。

We should improve urban-rural environmental infrastructure and continue to implement major projects for protecting and restoring key ecosystems.

加强住房保障体系建设，支持刚性和改善性住房需求，解决好新市民、青年人等住房问题。

We should improve the housing support system, support people in buying their first homes or improving their housing situation, and help resolve the housing problems of new urban residents and young people.

第十一章　翻译腔应对策略

翻译腔首先由尤金·奈达等在其著作《翻译理论与实践》(2004)中提出，英文为 translationese，即将源语的语言形式、表达方式、句法结构机械地移植到译语中，由此形成一种不土不洋、不符合译语表达习惯的语言混合体。马克·沙特尔沃思(Mark Shuttleworth)等人(2005)将 translationese 定义为"因明显依赖源语的词语和结构而让人觉得不自然、费解甚至可笑的目标语用法"。陆谷孙(2007)将 translationese 翻译为"翻译腔"，即"表达不流畅、不地道的翻译文体"和"佶屈聱牙的翻译语言"，也有学者将 translationese 翻译为"翻译体"。本书采用"翻译腔"这一术语。

科技翻译中的翻译腔可能由以下原因产生：

(1) 译者英语熟练程度不高；

(2) 译者不够熟悉汉英两种语言之间的差异及其背后的中西思维模式差异；

(3) 译者忽略中西方文化差异；

(4) 译者对英语写作原则、翻译技巧等知识不够了解或理解有偏差；

(5) 译者崇尚翻译腔；

(6) 译者对源语文本理解不够准确；

(7) 译者缺乏翻译实践经验。

针对以上原因，可通过以下方法改善翻译腔：

(1) 加强英语有效输入，提高英语语言能力；

(2) 研读汉英语言对比书籍，了解两种语言之间的差异；

(3) 了解中西文化差异，并有意识地将其体现在翻译中；

(4) 广泛阅读英语写作、翻译技巧相关书籍，积累丰富的翻译知识；

(5) 对翻译腔有正确认识，不能以翻译腔为荣；

(6) 有一定的科技常识，正确理解源语文本；

(7) 大量实践，提高翻译能力。

我国学者主要研究外译汉中的翻译腔，本书参考中介语理论、汉英语言对比研究、英语为母语国家权威写作著作等，总结科技文本汉译英中翻译腔的常见表现并提出应对策略。

第一节 直译与意译相结合

直译(literal translation)是尽可能地贴近原文内容与形式的翻译，是相对于意译(free Translation)而言的。英文中还有一个类似于 literal translation 的表达，即 word for word translation，学界常将其翻译为"逐词(翻)译"，也有学者讨论直译和逐词(翻)译的区别，认为直译考虑目标语的语法规范，对原文形式作必要的调整，因而译文比较通顺，读者能够看懂，而逐词(翻)译不顾目标语的语法规范和表达习惯，一味追求形式对等，译文晦涩生硬，难以理解。方梦之(2018)将 literal translation 翻译为"逐字译"，但他同时也使用"直译"这一术语。

直译法是一种常见的翻译方法，能够传达原文意义，体现原文风格等。然而，很多译者在翻译中生硬地使用直译法，导致译文不通顺，不符合目标语表达习惯，即翻译腔严重，使信息传递效果大打折扣。

在翻译实践中，应将直译、意译相结合，若原文的思想内容与译文的表达形式一致，可使用直译法，若有矛盾，则采用意译法。

【经典赏析】

原文：中国铁塔通过"铁塔＋5G＋AI"，将超过 20 万座"通信塔"变为"数字塔"。

译文：China Tower has made more than 200,000 communication towers smarter by equipping them with 5G and artificial intelligence.

赏析：本句采用直译和意译相结合的方法。"通信塔"采用直译法译为 communication towers，而将"数字塔"意译为 smarter，将"铁塔＋5G＋AI"这一公式化结构用词语表达。整个译文通顺流畅，可读性强。

【案例解析】

◇ 例1

原文：我国将持续优化基础设施布局，提升 5G、千兆光网等高质量网络覆盖深度广度，打造一批 5G 工厂。

原译：China will continuously optimize the infrastructure layout, improve the depth and width of the coverage of high-quality networks such as 5G and gigabit optical networks, and build a batch of 5G factories.

解析：原译采用直译法，结构和汉语高度一致，"深度广度"也忠实地翻译为

depth and width。仔细阅读原文可知，"持续优化基础设施布局"可理解为"提升 5G、千兆光网等高质量网络覆盖深度广度"的手段，翻译为 by doing…，使句子结构更符合英语表达习惯。"覆盖"本身就隐含了"广度"，"深度"和"广度"常搭配使用，构成四字短语，没有必要翻译出来。

译文：China will improve the coverage of high-quality networks such as 5G and gigabit optical networks by continuously optimizing the infrastructure layout, and build a batch of 5G factories.

◇ 例 2

原文：Cloud Object Storage 能够为组织提供必要的灵活性、可扩展性和简便性支持，帮助他们在混合云环境下存储、管理和访问眼下迅速增长的非结构数据。

原译：Cloud Object Storage provides organizations with flexible, scalable and simple support and helps them to store, manage and visit today's rapidly growing unstructured data in a mixed cloud environment.

解析：原译完全按照汉语语序和结构，动词为 and 连接的 provides…and helps…，略显冗长；"灵活性、可扩展性和简便性"为"支持"的具体内容，而非"支持"的定语，因此"支持"不必翻译出来；将"访问"和"混合"分别译为 visit 和 mixed 不够专业。基于以上分析，可将句子结构稍作调整，将第二个动词短语调整为状语，使句子结构更简洁、更符合英语表达习惯；将"灵活性、可扩展性和简便性"译为名词，省去对"支持"的翻译；按照计算机术语，将"访问"和"混合"分别译为 access 和 hybrid。

译文：Cloud Object Storage provides organizations with the flexibility, scalability and simplicity required to store, manage and access today's rapidly growing unstructured data in a hybrid cloud environment.

◇ 例 3

原文：必须指出的是，高学历或高薪职位并不能阻止你的饭碗被人工智能抢走。

原译：It's important to note that high academic qualifications or high-paying positions cannot prevent AI from robbing you of your rice bowl.

解析：原译完全直译，将"高学历""饭碗""抢走"等词按照字面意思处理，使译文冗长、晦涩。可将"高学历"理解为"高等教育""饭碗"理解为"工作",甚至"阻止"都可灵活翻译。

译文1:It's important to note that an advanced education or a high-paying position is not a defense against AI takeover.

译文2:It's important to note that an advanced education or a high-paying position is not a defense against job loss to AI.

译文3:It's important to note that an advanced education or a high-paying position cannot prevent AI from usurping human jobs.

◇ 例4

原文:伴随企业数字化进程的发展,数据的产生、采集、传输、处理、存储和管理都发生了巨大改变,体现出数据海量增长、数据类型丰富、数据的传输与处理要求攀升、数据管理复杂度增加等特征,企业需要构建未来数字化基础架构以应对挑战。

原译:With the development of the enterprise digitalization process, the generation, collection, transmission, processing, storage, and management of data have undergone huge changes, embodying such features as data increasing greatly, data types being rich, requirements for data transmission and processing becoming higher, and data management becoming increasingly complex, and enterprises need to establish a future digital infrastructure to face the challenges.

解析:原译完全按照原文的措辞和结构,行文不够简洁,如 digitalization 和 process 意思有重叠,用动名词复合结构 data increasing greatly, data types being rich, requirements for data transmission and processing becoming higher, and data management becoming increasingly complex 描述 features 等;全句过长,可读性不强;"未来"简单翻译为 future,不符合作者意图。基于以上分析,可将不必要出现的词删掉;将动名词复合结构改为名词短语;将原句拆分为三句话,便于读者理解;将"未来"翻译为 future-facing,使表达更接近作者意图。

译文:As enterprises progress in digital transformation, data generation, collection, transmission, processing, storage, and management have undergone huge changes. These changes have such features as explosive data growth, diverse data types, higher data transmission and processing requirements, and the increasing complexity of data management. To address these challenges, enterprises need a future-facing digital infrastructure.

◇ 例 5

原文：全闪存数据中心是指 90%以上的存储容量需由固态硬盘提供(包括外置存储系统与服务器内置存储)的数据中心，且同时具备高密度、高可靠、低延迟、低能耗等特征，可承载企业新兴业务和应用，可作为帮助企业最大化实现数据创新的数据中心。

原译：An all-flash data center refers to a data center 90% or more of whose storage capacity (covering external storage systems and built-in storage of servers) needs to be provided by SSDs, and at the same time it has the features of high density, high reliability, low latency, and low energy dissipation, carries the emerging businesses and applications of enterprises, and helps enterprises realize data innovation to the greatest extent.

解析：原译严格按照汉语语序和措辞，将"90%以上的存储容量需由固态硬盘提供的数据中心"译为 90% or more of whose storage capacity needs to be provided by SSDs，语法虽正确，但难以理解，且重要信息 SSDs 出现在句末，未能凸显其重要性；句中 and 较多，译文冗长拖沓；句子过长，可读性不强；有些短语翻译过于生硬，如将"90%以上"译为 90% or more，将"最大化实现"译为 realize… to the greatest extent。此外，将"业务"译为 businesses，使读者对 businesses 和 enterprises 产生混淆。基于以上分析，将"90%以上的存储容量需由固态硬盘提供的数据中心"灵活翻译；将句子拆分，避免过多 and 的出现，并使句子长短适中，易于理解；短语灵活翻译，使译文自然、简洁。

译文：An all-flash data center refers to a data center that uses SSDs for at least 90% of its storage capacity (covering external storage systems and built-in storage of servers) while delivering high density and reliability, low latency, and energy efficiency. The all-flash data center helps enterprises harness the power of emerging technologies and applications and maximize data innovation.

练习 *practice*

1. 简答题。

(1) 产生翻译腔的原因有哪些?

(2) 直译的优点和缺点分别有哪些?

2. 赏析下列翻译，指出译文的优点。

(1) 国家知识产权局 12 月 28 日发布的报告显示，中国发明专利产业化率近 5 年稳步提高。2022 年，中国有效发明专利产业化率为 36.7%，其中，企业发明专利产业化率为 48.1%。

The industrialization transfer rate of China's effective patents in 2022 reached a five-year high of 36.7 percent, an annual survey said on Dec 28. Issued by the National Intellectual Property Administration, the survey said the patent industrialization rate for enterprise patentees was 48.1 percent this year.

(2) 据国家航天局消息，11 月 21 日，国家遥感数据与应用服务平台发布暨国家航天局卫星数据与应用国际合作中心揭牌仪式在海南海口举办。国家遥感数据与应用服务平台正式开通，旨在构建国家级、综合性遥感资源共享与应用服务体系，进一步推进遥感卫星数据资源共建共用共享，带动我国遥感综合应用能力迈上新台阶。

The China National Space Administration (CNSA) on Monday launched a national platform for remote sensing data and application services in Haikou, capital city of the southern island province of Hainan. Also on Monday, the CNSA held an opening ceremony for its international satellite data and application cooperation center in Haikou. China has launched the platform to build a national and comprehensive remote sensing resource sharing and application services system. And the platform will promote the contribution and sharing of resources further.

(3) 第三平台技术(云计算、大数据与分析、移动、社交、人工智能、下一代安全、物联网以及区块链、量子计算等下一代科技)的落地与进步，促进了生产生活中广泛的创新，催生了新的经济模式，创造新的业务形态，带来新的商业价值。

The emergence and progress of 3rd Platform technologies and services has inspired a range of innovations that generate new economic patterns, create new business forms, and bring new commercial value. The 3rd Platform technologies typically include cloud computing, big data and analytics, mobility, social business, artificial intelligence technologies, next-generation security, the Internet of Things (IoT), and other next-generation technologies such as blockchain and quantum computing.

(4) 随着机器视觉技术以及人工智能(AI)视频图像分析应用的兴起，越来越多的场景将会采用智能的摄像系统，到 2025 年，全球视频监控摄像头的市场规模将达到 440 亿美元，智能视频图像系统将会产生大量的数据，尤其是非结构化数据。

With the rise of machine vision technologies and AI video image analytics applications, intelligent camera systems will be adopted in more scenarios. The world-wide video surveillance camera market will grow to 44 billion USD by 2025. Intelligent video image systems will generate a large amount of data, a significant proportion of which will be unstructured.

(5) 当前的 NVMe over Fabrics 主要有两类主流的方式，包括 FC-NVMe 和 NVMe over RoCE，前者被采用的主要原因是 FC 网络是目前存储网络的主流选择，其在稳定性方面具有一定的优势，NVMe over RoCE 则通过更加开放的以太网，以及带宽速度的快速演进等优势，展示出更具发展潜力的态势。

Currently, there are two mainstream forms of NVMe over Fabrics: FC-NVMe and NVMe over RoCE. The former is adopted in many mainstream storage networks because FC networks deliver certain advantages in stability. NVMe over RoCE is considered to be a more promising option with advantages in more open Ethernet and rapid improvement of bandwidth speeds.

3. 翻译下列句子，注意避免生硬直译。

(1) 随着智慧城市的建设，安防需求与日俱增，跨模态行人重识别成为学术界、工业界的热点研究方向。

(2) 全闪存具有更高的可管理性和可维护性，同时，固态硬盘的尺寸正变得更加灵活，通常可提供多种长度、宽度和高度选项。

(3) 在新的业务需求下，内存驱动的基础架构是未来发展趋势之一。

(4) 现代查询优化器需要对表中的数据结构和数据构成进行详细统计，从而针对如何执行复杂查询做出"最佳"决策。

(5) IT 基础设施容易发生服务器故障、磁盘崩溃或存储损坏、站点中断和人为错误等各种故障，这些故障会造成成本高昂的计划外停机。

第二节　形合与意合相结合

所谓形合与意合，主要指分句与分句之间起连接作用的成分是保留还是省略的一个句法问题。汉语和英语属于不同语系，汉语重意合，以意统形，以神驭形，行文铺排疏放，结构不求完整，句子常缺主干，核心动词谓语往往不甚明显，不易形成信息焦点。同时汉语句式简短，多表现为自然的线性序列，通常情况下，句中各成分的相互结合大多依靠语义的贯通、语境的映衬和词序的排列，较少使用连接词语。相比之下，英语重形合，以形显义，连接词在大多数情形下不可缺少(黄忠廉，2022)。流水句为汉语的一大特色，由小句与小句构成，语义关系松散，无连接词。在翻译此类句子时，要注意汉英两种语言的区别，形合、意合相结合，避免译文句子结构松散，突出英语的特点，提高译文的可读性。可通过标点符号、词形变化、句子重构、增添衔接词等方式实现意合向形合的转变。

【经典赏析】

原文：斯特林热电转换试验装置也顺利完成在轨试验，斯特林热电转换系统能将热能高效转化为电能，在未来载人月球及深空探测任务中，具有广阔的应用前景。

译文：The Stirling thermoelectric convertor has also completed its in-orbit test. Capable of converting thermal energy into electricity with relatively high efficiency and power density, the convertor is expected to be used in future manned lunar missions and deep-space exploration

赏析：原文汉语为流水句，译者首先使用句号将其拆分为两句话。重构拆分之后的第二句话，将其中第一个小句处理为形容词短语，置于句首作状语，对句子主语作补充说明，这样处理符合英语表达习惯，译文可读性强。

【案例解析】

◇ 例 1

原文：到 2035 年，可再生能源制氢在终端能源消费中的比重明显提升，对我国能源绿色转型发展起到重要支撑作用。

原译：By 2035, the proportion of hydrogen produced from renewable energy in terminal energy consumption will increase significantly, this will play an important supporting role in China's green energy transformation.

解析：原文为典型的汉语流水句，译者受原文影响，将其翻译为两个分句，但之间无连接词，不符合英语语法规范。为避免此类语法错误，可采取以下几种方法：① 将两个分句之间的逗号改为句号；② 在逗号之后添加 and；③ 将其中一个分句处理为短语或从句。其中方法①使两句话过于单薄，方法②使行文不够简洁、紧凑。根据语境，建议本句翻译采用方法③。

译文：By 2035, the proportion of hydrogen produced from renewable energy in terminal energy consumption will increase significantly, which will play an important supporting role in China's green energy transformation.

◇ 例2

原文：未来企业就是坚定走数字化道路的组织，数字化运营实现规模化，创新速度比传统企业高出一个数量级。

原译：The future enterprises are those organizations that firmly follow the digital path, digital operations achieve scale, and the innovation speed is a magnitude greater than that of traditional business.

解析：原译未能将小句之间的逻辑关系呈现出来，可读性不强。仔细阅读可发现，第二和第三小句是对第一小句的进一步解释说明，可用非限定性定语从句表达。

译文：The future enterprises are those organizations that firmly follow the digital path, where digital operations achieve scale, and the innovation speed is a magnitude greater than that of traditional business.

◇ 例3

原文：只有通过全球范围内技术厂商、供应商、客户以及标准、政策与法律制定者之间的合作，才能在应对该挑战上取得积极显著的成效。我们将共享知识和经验，务实合作，共同努力，减少技术被滥用所导致的不可预期风险。

原译：Only through global collaboration between vendors, service providers, and customers, as well as industry standards bodies, and policy and lawmakers, will we be able to effectively address these challenges and deliver positive measurable results. We should share knowledge and experience, be both pragmatic and cooperative, work together, and reduce unforeseen cloud security risks rooted in the misuse and abuse of technologies.

解析：原文两句话之间无衔接词，逻辑关系隐含在语义中，第二句话包含多个小句。原译受汉语结构影响，两句话之间语义结构松散。第二句话包含小句过多，略显冗长。建议在两句话之间添加衔接手段，并将第二句话根据语义进行拆分，将逻辑关系显性化，增强可读性。

译文：Only through global collaboration between vendors, service providers, and customers, as well as industry standards bodies, and policy and lawmakers, will we be able to effectively address these challenges and deliver positive measurable results. Along the way, we are committed to sharing our knowledge and experience, as well as staying both pragmatic and cooperative. By joining forces, we can successfully handle unforeseen cloud security risks rooted in the misuse and abuse of technologies.

◇ 例 4

原文：我国基础研究投入从 2015 年的 716 亿元增长到 2019 年的 1335.6 亿元，年均增幅达到 16.9%，大大高于全社会研发投入的增幅。

原译：China's investment in basic research increased from 71.6 billion yuan in 2015 to 133.56 billion yuan in 2019, its average annual growth rate reaches 16.9 percent, and this is much higher than that of the spending on research and development.

解析：原文为典型的流水句，包含三个小句。原译完全按照汉语结构，未考虑汉语和英语语言的区别。可将后两个分句处理为状语，将三个小句整合为一句话，结构紧凑，逻辑清晰。

译文：China's investment in basic research increased from 71.6 billion yuan in 2015 to 133.56 billion yuan in 2019, with an average annual growth rate of 16.9 percent, much higher than that of the spending on research and development.

◇ 例 5

原文：虚拟机对内存的访问都会经过虚拟化层的地址转换，保证每个虚拟机只能访问到分配给它的物理内存。

原译：Virtual machines' access to memory resources entails address translation at the virtualization layer, and this ensures that each virtual machine can access only the physical memory resources that have been assigned to it.

解析：第二个小句与第一个小句为递进关系，作进一步解释说明，翻译时处理
　　　为非限定性关系分句，可使行文更简洁，结构更紧凑，更符合英语习惯.

译文：Virtual machines' access to memory resources entails address translation at the
　　　virtualization layer, which ensures that each virtual machine can access only
　　　the physical memory resources that have been assigned to it.

练 习
practice

1. 赏析下列翻译，指出汉语的意合如何转换为英语的形合。

(1) 该基站的增强系统由国产的北斗卫星导航系统提供，该系统可将动态
定位精度提升至厘米级，静态定位精度提升至毫米级。

Provided by the China-made Beidou Navigation Satellite System, the
station's augmentation system can increase positioning precision to the
centimeter-level dynamically and the millimeter-level statically.

(2) 我国有望在 2025 年前后达到碳排放峰值，提前五年完成 2030 年《巴黎
协定》目标。

The peak in China's carbon emissions is estimated to be hit around 2025,
five years ahead of its 2030 Paris Agreement target.

(3) 我国将依靠企业、市场化机制和其他资源推动人工智能开放创新平台
建设，能够产生全新有效的服务从而促进社会经济增长。

China will rely on enterprises, market mechanisms and other resources to
build more open and innovative artificial intelligence platforms capable of
producing new and effective services to bolster socioeconomic growth.

(4) 随着数字经济快速发展，我国将采取更多措施促进互联网行业持续健
康发展，推动数字经济与实体经济深度融合。

While the digital economy is expanding fast, China will take more
measures to promote the sustainable and healthy development of the
internet sector for in-depth integration of digital and real economies.

(5) 中国制造业要走向高质量发展，要强化国家战略引领，引导创新要素
更多投向核心技术攻关，大力营造公平竞争的市场环境。

To pursue high-quality development in the manufacturing sector, it is of
great significance to step up strategic guidance at a national level to enable
innovation to flow into the research of core technologies and create a fair

market environment.

2. 翻译下列句子。

(1) 该技术逐渐成熟，规模化落地实践已经具备了现实基础。

(2) 量子力学是物理学的一个分支，描述了物质和能量在最微小尺度上的奇怪行为。

(3) "3CS" 体系的基础理念是基于云服务各业务模块的流程，划分相对应的安全控制领域，使安全控制要求得以嵌入到云服务管理流程中，同步确保安全管理责任清晰明确、可度量、可追溯。

(4) 华为云将基于安全责任共担模型，持续主动地构建并提升包括物理环境、网络、平台等各层基础设施与云服务的安全合规能力，全面保障用户业务与数据的安全与合规。

(5) 这使得内存空间成为一个巨大的攻击面，攻击者可以利用实时微控制器系统中的内存破坏漏洞成功地破坏任何软件模块。

(6) 第七届中国工业大奖获奖名单 3 月 19 日公布，19 家企业、19 个项目获得中国工业大奖。

(7) 三维打印也称为增材制造，是一个技术系列，采用虚拟的计算机辅助设计(CAD)模型，通过连续创建层来创建一个物理对象。

(8) 中国电信持续开展全光网络建设，目前千兆光网已覆盖 300 多个城市。

(9) 中国人工智能大模型研发正在蓬勃发展，各种技术路线都在并行突破。

(10) 数据的共享、流通、交易和应用被视为一种新型的生产要素，是促进我国数字化发展的关键。

第三节 动词与名词化结构相结合

如前所述，名词化是从其他词类变为名词的过程，特别是动词和形容词，其结果是将动作过程或性状等进行"事物化""固化"或"静态化"处理。名词化结构多用于学术语境，例如：

从传感器引线到现场配线的转变通常是在连接到传感器的连接头上完成的。

The transition from the sensor lead wires to the field wiring is typically done in a connection head attached to the sensor.

但过多使用名词化结构，可能会导致句子繁杂沉闷、晦涩难懂，而且使动作显得苍白无力，重点不突出，例如：

IT 系统安全涉及通过对来自贵企业内外部的非法访问进行阻止、检测和响应来保护系统和信息。

IT system security involves the protection of systems and information through prevention and detection of and response to improper access from within and outside your enterprise.

若将此句后半部分改为动词的-ing 形式，则语言更简洁、生动。

IT system security involves the protection of systems and information by preventing, detecting, and responding to improper access from within and outside your enterprise.

近年来，越来越多的文体学家将焦点放在文本的交际功能上，主张多用动词，以增强文本的可读性。以英语为母语的学术写作学者也大多提出，写作要经济、清晰、明了，在可能的情况下，尽量使用动词表达动作。

实际上，动词与名词化结构相结合、具体与抽象相结合合才是最佳选择。科技文本为正式文体，恰当使用名词化结构可使语体更正式、更客观、更简洁、更专业，而且可以提高语篇内部的衔接性。但同时要避免滥用名词化，若名词化结构不能取得以上效果，则最好使用动词形式。

【经典赏析】

原文：碳中和，即通过能效提升和能源替代将人为活动排放的二氧化碳减至最低程度，然后通过森林碳汇或捕集等其他方式抵消二氧化碳的排放，实现源与汇的平衡。

译文：Carbon neutrality refers to achieving a balance between carbon emission and carbon sinks by enhancing energy efficiency and energy replacement to minimize the carbon dioxide emission by human activities and then offset such emission with forest carbon sinks or ways like sequestration.

赏析：将"人为活动排放的二氧化碳"译为 the carbon dioxide emission by human activities，将"能源替代"译为 energy replacement，使用名词化使行文简洁、正式。将"实现源与汇的平衡"译为 achieving a balance between carbon emission and carbon sinks，将"通过能效提升和能源替代"译为 by enhancing energy efficiency and energy replacement，使用动词 achieving 和 enhancing，直接、生动，比使用名词化结构 the achievement of 和 the enhancement of 更简洁。译文将动词和名词化结构相结合，取得了最佳交流效果。

【案例解析】

◇ 例1

原文：在以下几段，我们将分析全方位和自适应超光谱传感器的性能。

原译：In the following paragraphs, we will make an analysis of the performance of full and adaptive hyperspectral sensors.

解析：原译 make an analysis of 不够简洁，可改为动词 analyze。当出现没有实际意义的"动词＋名词化结构＋介词/that"时，可考虑直接用动词替换，如 make an investigation of→investigate，conduct a study of→study，put forward a proposal→propose，put emphasis on→emphasize 等。

译文：In the following paragraphs, we will analyze the performance of full and adaptive hyperspectral sensors.

◇ 例2

原文：非法访问会导致信息变更、损毁、盗用或滥用，或导致你的系统被破坏或滥用。

原译：Improper access can result in the alteration, destruction, misappropriation or misuse of information or can result in damage to or misuse of your systems.

解析：原译过多使用名词化结构，行文略显晦涩，the alteration, destruction, misappropriation or misuse of information 的名词化结构头重脚轻，读者不易抓住重点。直接使用"名词＋动词-ing 形式"的动名词复合结构直观、易懂，即 information being altered, destroyed, misappropriated or misused 作介词 in 的宾语。并列动词短语 or can result in damage to or misuse of your systems 保留名词化结构，避免前文在结构、用词上的重复。动词和名词化结构相结合，才会使语言表达更多样、更生动。

译文：Improper access can result in information being altered, destroyed, misappropriated or misused or can result in damage to or misuse of your systems.

◇ 例3

原文：最后，我们讨论了研究问题，并提出了进一步改进的未来方向。

原译：We conclude with a discussion of research issues and a suggestion of future directions for further improvement.

解析：将"最后"译为 conclude，用词正式，符合学术英语文体风格，但 with 后的名词化结构 a discussion of… and a suggestion of…不够简洁。可将 with 改为 by，相应的名词化结构改为动词-ing 形式。

译文：We conclude by discussing research issues and suggesting future directions for further improvement.

◇ 例4

原文：训练深度模型的一个关键挑战来自这一点：训练样本数量与可学习参数的数量相比极其有限。

原译：A critical challenge in the training of deep models arises from the limited number of training samples compared with the number of learnable parameters.

解析：原译 the training of 为名词化结构，略显冗长，可用动词-ing 形式 training 代替，使行文更简洁。

译文：A critical challenge in training deep models arises from the limited number of training samples compared with the number of learnable parameters.

◇ 例5

原文：在这里，我们报道了通过牛顿流体和粘弹性流体的微流体共流来分离小于 3μm 的微粒，这种分离取决于微粒大小，可利用这两种流体之间的界面效应取得。

原译：Here we report using a microfluidic co-flow of Newtonian and viscoelastic fluids to separate microparticles smaller than 3μm. The separation depends on the size of microparticles and can be achieved through the use of the interfacial effect between these two types of fluids.

解析：原译不够简洁。原文出现两次"分离"，翻译时可考虑使用名词化结构将两部分整合。最后一部分可考虑使用动词-ing 形式，避免名词化结构过多。

译文：Here we report a microfluidic co-flow of Newtonian and viscoelastic fluids for size-dependent separation of microparticles smaller than 3μm by utilizing the interfacial effect between these two types of fluids.

练习 *practice*

1. 赏析下列翻译，指出动词和名词化结构如何相结合。

(1) 这种焦虑导致人们回避日常情景，从而对身心健康产生不良影响。

This anxiety leads to withdrawal from everyday situations, which adversely affects both mental and physical health.

(2) 例如，在红外测量时，需要对催化剂的发射率有可靠的了解才能获得准确的表面温度测量值。

For instance, in the case of infrared measurements, it is necessary to understand the catalyst's emissivity reliably to obtain accurate surface temperature measurements.

(3) 虽然抵御这种对抗性扰动的研究仍处于起步阶段，但事实证明，JPEG 可以自然消除这些难以察觉的扰动，从而恢复图像的原始分类。

While the research in defending against such adversarial perturbations is still in its infancy, it has been shown that JPEG naturally removes some of these imperceptible perturbations, thereby restoring the original classification of the image.

(4) 这在这种应用中尤其不可取，因为它可能导致在跟踪 IOP 随时间变化时出现误差，以及对疾病进展监测不准确。

It is especially undesirable in this application, as it could lead to errors in tracking changes in IOP over time and inaccurate monitoring of the progress of diseases such as glaucoma.

(5) 数据包处理涉及传入数据包的解析与分类以及高吞吐量(25～400 Gb/s)的庞大流量(1～20 MB)的处理。

Packet processing involves parsing and classification of incoming packets and processing a large number of flows (1-20 MB) at high throughputs (25-400 Gb/s).

2. 翻译下列句子，注意不要过度使用名词化结构。

(1) 现在我们可以对个体的因果效应进行定义。

(2) 我们已经成功地在新西兰白兔中植入了干涉式压力传感器，并获得了对兔子一只眼睛的眼内压力直接测量的数据。

(3) 为了对这种差异进行进一步研究，我们采用了另一种方法，使用谐振频率和位移测量来测量 SiN 薄膜中的残余应力。

(4) 由于需要大量计算资源，因此无法实现用 CPU 核心以这样的吞吐量进行流量处理。

(5) 在各种基于微控制器的嵌入式系统中，实时微控制器系统的设计是为了满足实时进程的时限约束。

第四节　主动语态与被动语态相结合

科技英语主要阐述、表达客观事物的本质特性、变化过程以及与其他事物之间的联系，因此被动语态成为科技英语文体的显著特点。据统计，科技英语文体中，约有 25%的动词使用被动语态。被动语态可以使语言更客观、更规范、更简洁。此外，使用被动语态，将所要讨论的对象放在主语的突出位置，也可以将所要论证及说明的主旨置于更突出的地位。

近几十年来，也有学者主张在学术论文或报告中使用人称代词，凸显人际意义，以加强学术交流中的亲和力。英语国家权威写作专家威廉·斯特伦克(William Strunk)等人也提倡多用主动语态，认为主动语态表达直接、生动。国际标准化组织 1982 年颁布的 ISO 5966-1982(e)Documentation-Presentation of Scientific and Technical Reports 也提倡在摘要中使用主动语态。

本书认为，在翻译实践中，应该根据语境、作者意图、句子结构等多种因素，灵活使用两种语态，提升交流效果。过度使用被动语态，可能使语言不够自然，因此，若非为了特殊意图(如强调宾语、取得句子平衡、与下文衔接等)，一般情况下，若句子主语明确，使用主动语态为好。但若句子主语为泛指的名词或人称代词，或主语不够明确，则可考虑使用被动语态以强调宾语，使文体更正式。

【经典赏析】

原文：近日，中共中央、国务院印发了《数字中国建设整体布局规划》。《规划》指出，建设数字中国是数字时代推进中国式现代化的重要引擎，是构筑国家竞争新优势的有力支撑。

译文：China has rolled out a plan for the overall layout of the country's digital development. Building a digital China is important for the advancement of Chinese modernization in the digital era, and provides solid support for the development of new advantages in the country's competitiveness, according to the plan, which was jointly released by the Communist Party of China Central Committee and the State Council.

赏析：第一句话主语明确，使用主动语态。但若直译，主语过长，则《规划》相关信息无法凸显，故将主语信息拆分，先以"中国"为第一句话主语，而将"中共中央、国务院"后置至第二句。第二句话首先描述规划内容，使重点信息突出，再以状语 according to the plan 补全信息，最后以被动语态引出"中共中央、国务院"，流畅自然。

【案例解析】

◇ 例1

原文：本文介绍了医学成像领域的计算机辅助图像分析。

原译：Computer-assisted analysis of images in the field of medical imaging is covered in this paper.

解析：本句主语明确，为"本文"，但译者刻意使用被动语态，导致行文烦琐，且头重脚轻。使用主动语态可克服这一问题。

译文：This paper covers computer-assisted analysis of images in the field of medical imaging.

◇ 例2

原文：我们讨论了目前对液-液相分离(LLPS)的理解，总结了其生理功能，并进一步地描述了 LLPS 在人类疾病发展中的作用。此外，我们还回顾了最近所开发的研究 LLPS 的方法。

原译：Our current understanding of LLPS is described and its physiological functions are summarized. The role of LLPS in the development of human diseases is further described. Additionally, the recently developed methods for studying LLPS are reviewed.

解析：原译使用被动语态，第二句话出现头重脚轻的现象。此外，第一句已出现人称代词"我们"，若使用主动语态以人称代词 we 为主语，逻辑上更严密，并且可以拉近作者和读者的距离，使语言更生动。

译文：We present our current understanding of LLPS and summarize its physiological functions. We further describe the role of LLPS in the development of human diseases. Additionally, we review the recently developed methods for studying LLPS.

◇ 例3

原文：我们介绍了深度学习方法的基本原理，并回顾了它们在图像注册、解剖和细胞结构检测、组织分割、计算机辅助疾病诊断和预后等方面的成功经验。最后，我们讨论了研究问题并提出进一步改进的方向。

原译：The fundamentals of deep learning methods are introduced and their successes in image registration, detection of anatomical and cellular structures, tissue segmentation, computer-aided disease diagnosis and prognosis, and so on are reviewed. Finally, we discuss research issues and suggest directions for further improvement.

解析：原译第一句使用被动语态，使得第二个分句的主语过长。第二句和第一句结构相似，却使用人称代词 we 作主语，切换主语导致行文不够流畅。为解决以上问题，可采用主动语态。

译文：We introduce the fundamentals of deep learning methods and review their successes in image registration, detection of anatomical and cellular structures, tissue segmentation, computer-aided disease diagnosis and prognosis, and so on. We conclude by discussing research issues and suggesting directions for further improvement.

◇ 例4

原文：通过利用空间域以及高光谱域，我们减少了误报率，而没有使召回率大幅下降。

原译：By exploiting the spatial domain as well as the hyperspectral domain, the false alarm rates were reduced without degrading the recall rates dramatically.

解析：原译使用被动语态，使得 exploiting 和 degrading 没有逻辑主语，成为悬垂结构。将句子改为主动语态可解决该问题。

译文：By exploiting the spatial domain as well as the hyperspectral domain, we reduced the false alarm rates without degrading the recall rates dramatically.

◇ 例5

原文：本文的其余部分安排如下。第二节描述了背景。第三节定义了我们的威胁模型。第四节详细介绍了我们的模型的设计。第五节介绍了我们的模型与相关研究中的模型的区别。第六节是本文的结论。

原译：The rest of this paper is organized as follows. Section II describes the background. Our threat model is defined in Section III. Section IV presents the design of our model in detail. The differences between our model and those mentioned in related work are shown in Section V. Section VI concludes this paper.

解析：译者交替使用主动语态和被动语态，似乎是为了让语言更生动。但频繁更换主语可能破坏行文的流畅。

译文：The rest of this paper is organized as follows. Section II describes the background. Section III defines our threat model. Section IV presents the design of our model in detail. Section V shows the differences between our model and those mentioned in related work. Section VI concludes this paper.

练习
practice

1. 阅读下文，分析具体语境中使用主动语态或被动语态的原因。

Real-time microcontrollers have been widely adopted in cyber-physical systems that require both real-time and security guarantees. Unfortunately, security is sometimes traded for real-time performance in such systems. Notably, memory isolation, which is one of the most established security features in modern computer systems, is typically not available in many real-time microcontroller systems due to its negative impacts on performance and violation of real-time constraints. As such, the memory space of these systems has created an open, monolithic attack surface that attackers can target to subvert the entire systems. In this paper, we present MINION, a security architecture that intends to virtually partition the memory space and enforce memory access control of a real-time microcontroller. MINION can automatically identify the reachable memory regions of real-time processes through off-line static analysis on the system's firmware and conduct run-time memory access control through hardware-based enforcement. Our evaluation results demonstrate that, by significantly reducing the memory space that each process can access, MINION can effectively protect a microcontroller from various attacks that were previously viable. In addition, unlike conventional memory isolation mechanisms that might incur substantial performance overhead, the lightweight design of MINION is able to maintain the real-time properties of the microcontroller.

2. 赏析下列翻译，指出译文的优点。

(1) 本周二 15 时 37 分，我国在太原卫星发射中心使用长征四号乙运载火箭，成功将对地观测卫星高分十一号 04 星发射升空，卫星顺利进入预定轨道。

China sent a new Earth observation satellite into space from the Taiyuan Satellite Launch Center in north China's Shanxi Province at 3:37 p.m. Tuesday (Beijing Time). The satellite, Gaofen-11 04, was launched aboard a Long March-4B carrier rocket and entered its planned orbit successfully.

(2) 近日，中共中央办公厅、国务院办公厅印发了《关于新时代进一步加强科学技术普及工作的意见》(以下简称《意见》)。《意见》提出发展目标，到 2025 年，科普公共服务覆盖率和科研人员科普参与率显著提高，公民具备科学素质比例超过 15%，全社会热爱科学、崇尚创新的氛围更加浓厚。

China has issued a guideline on facilitating the popularization of science and technology. Released by the general offices of the Communist Party of China Central Committee and the State Council, the guideline sets specific targets for the popularization of science and technology. By 2025, public services for science popularization will be significantly expanded, more researchers will play an active role in spreading scientific knowledge, the proportion of citizens with scientific literacy will exceed 15 percent, and a social climate that values science and innovation will be created, it says.

(3) 本综述结构如下。在第 2 节，我们解释了神经网络和深度模型的计算理论，并讨论了它们如何从数据中提取高级表征。在第 3 节，我们介绍了使用深度模型进行各种医学成像实际应用的研究。最后，在第 4 节，我们总结了研究趋势并提出了进一步改进的方向。

This review is organized as follows. In Section 2, we explain the computational theories of neural networks and deep models and discuss how they extract high-level representations from data. In Section 3, we introduce recent studies using deep models for different applications in medical imaging. Finally, in Section 4 we conclude by summarizing research trends and suggesting directions for further improvements.

3. 以适当语态翻译下列句子。

(1) 本文具体贡献如下：我们训练了迄今为止最大的神经网络之一。我们编写了高度优化的 2D 卷积 GPU 实现以及训练卷积神经网络内部的所有其他操作，并将其公开。

(2) 从低分辨率(LR)图像来估计其对应高分辨率(HR)图像是一项挑战性极强的工作，被称为超分辨率(SR)。SR 在计算机视觉研究领域受到广泛的关注并被大量使用。

(3) 不同于以前的工作，我们定义了一种新的感知损失。图 1 展示了一张极为逼真的示例图像，该图像是使用 4 倍的上采样系数进行超分辨率处理的。

(4) Dong 等人使用双三次插值对输入图像进行上采样，并端到端地训练了一个三层的全卷积网络以取得最佳 SR 性能。之后的研究表明使网络直接学习上采样滤波器可以进一步提高准确性和速度，使其性能更好。

(5) 我们的网络架构如图 2 所示。它包含 8 个学习层——5 个卷积层和 3 个全连接层。下面，我们将描述我们网络结构中的一些新奇或不寻常的特点。

第十二章　常见问题分析

译文质量通常是译者知识、能力、素养等的综合体现，译文中的常见问题大多可以溯源到译者身上。本章跳出语言维度，从译者角度分析译文问题，探究问题根源，以期启发译者，最终提高译文质量。

从知识层面上来说，译者最大的挑战是科技知识欠缺。科技翻译不同于普通文本的翻译，要求译者具有一定的科技知识，否则，译者可能无法准确理解原文，甚至误解原文，因而不能精确地再现原文信息。

从能力层面上来说，英语能力不足是一个普遍存在的问题。英语能力不足会导致书面表达的准确性、丰富性、连贯性和得体性下降。此外，译者面临的另一大问题是不能克服母语影响。译者依赖母语知识，将母语中的语言形式和意义挪用到译文里，就会出现翻译腔。

从素养层面来说，责任心和职业精神不够，仅满足于将汉语转换为英语，不能做到精益求精、追求细节等，也会使译文有失水准。

第一节　科技知识欠缺

本章所说的科技知识是指常识性知识，并非需要深入了解的专业领域知识。译者科技知识欠缺，可能会导致漏译、误译等问题。

为积累科技知识，译者应该养成终身学习的习惯，除专门学习相关专业知识外，在日常生活中应对科技知识抱有好奇心，时常浏览社交媒体、网页，翻阅报纸、杂志、书籍等。例如，《中国日报》双语新闻中的"每日一词"模块就是一个很好的学习渠道。该模块以汉英双语介绍当下的国家大事及热门话题，涵盖政治、经济、文化、民生、科技等话题，短小精悍，可读性强，翻译权威，并被中共中央宣传部主办的"学习强国"平台采纳。译者可关注其中的科技话题，既积累科技知识，也提高英语语言能力，还可学习翻译技巧。美国科普杂志《科学美国人》(https://www.scientificamerican.com/)以英语介绍科普知识，对译者也有类似帮助。总之，学习渠道不胜枚举，译者可根据自身需要进行选择。

积累知识的意义不仅仅在于知识本身。经常阅读科技知识，可以将知识内化为能力。译者接触的文本常常涉及各个专业领域，但因精力有限，不可能学尽所有领域。积累足够的科技知识，可以使译者拥有常识和专业敏感度，提高对科技文本的整体理解力，助力科技翻译。此外，大多数译者系统学习过人文学科知识，但理工科知识较为薄弱，学习科技知识可以帮助译者完善知识结构，提高译者的综合素养。积累知识的过程也是学习的过程，唯有保持终身学习的习惯，才能成为优秀的译者。

【经典赏析】

原文：本周二，国家重大科技基础设施高能同步辐射光源(HEPS)直线加速器满能量出束，成功加速第一束电子束。HEPS 由中国科学院高能所承担建设。

译文：The linear accelerator of China's high-energy synchrotron radiation light source successfully accelerated its first electron beam on Tuesday. The light source, the High Energy Photon Source (HEPS), is a major science infrastructure project in China built by the Institute of High Energy Physics under the Chinese Academy of Sciences.

赏析：根据专业知识和常识可以判断出"国家重大科技基础设施"是指"高能同步辐射光源"，而非"直线加速器"，从而厘清这几个名词之间的逻辑关系，避免误译。此外，可以判断出"满能量出束"和下文"成功加速第一束电子束"意思相近，因此可以省译部分文字，使译文简洁、专业。

【案例解析】

◇ **例1**

原文：锂电池不需要完全充电。事实上，最好不要完全充电，因为高压会给电池施加压力，从长远来看会对其造成损耗。

原译：Lithium batteries do not need to be fully charged. In fact, it is better not to fully charge them, because a high voltage stresses the battery and wears it out in the long run.

解析：根据专业常识，这里的"锂电池"是指锂离子电池 lithium-ion battery。手机和笔记本电脑使用的都是锂离子电池，真正的锂电池 lithium battery 由于危险性大，很少应用于日常电子产品。

译文：Lithium-ion batteries do not need to be fully charged. In fact, it is better not to fully charge them, because a high voltage stresses the battery and wears it out in the long run.

◇ 例 2

原文：跨系统协作业务的数据安全关键技术与应用

原译：Key Technologies and Applications of Collaborative Services Data Security for Cross-Systems

解析：根据专业常识可以判断出，该标题的意思为"跨系统协作业务的数据安全关键技术以及该技术的应用"，而非原译中的"跨系统协作业务的数据安全关键技术以及跨系统协作业务的数据安全的应用"。

译文：Data Security Key Technologies for Cross-System Collaborative Services and Their Applications

◇ 例 3

原文：Xilinx 提供了采用高级抽象语言 P4 进行数据包处理的工具，其可以实现数据包解析、分类、查找与数据包编辑功能。与基于 RTL 语言的实现相比，使用 P4 处理数据包可以在更高的抽象层实现。

原译：Xilinx offers tools for packet processing using the higher layer abstract language P4. They can analyze, classify, search and edit packets. Packet processing using P4 can be realized at a more abstract layer compared with implementation based on the RTL language.

解析：根据专业知识或下文可以判断出，"其可以实现数据包解析、分类、查找与数据包编辑功能"是描述 P4 而不是"工具"。"抽象""解析""实现"等词为专业术语，不能想当然地翻译。

译文：Xilinx offers tools for packet processing using the higher layer abstraction language P4, which enables packet parsing, classification, lookup, and packet editing functions. Packet processing using P4 can be implemented at a higher layer of abstraction than needed for the RTL language-based implementation.

◇ 例 4

原文：虚拟机的目的是要完整地模拟另一个环境，而容器的目的则是使应用程序能够移植。

原译：The purpose of a virtual machine is to fully emulate another environment, while the purpose of a container is to enable applications to transplant.

解析：由专业知识可以推断，虚拟机要模仿的环境不仅仅是"另一个"环境，而是隐含"不同环境"的意思，因此 another 不能充分再现原文信息。

根据语境，"移植"应有被动含义，且并非字面意思，查阅相关资料，得知容器具有可移植性(portability)，可以推测此处应译为 portable。

译文：The purpose of a virtual machine is to fully emulate a foreign environment, while the purpose of a container is to make applications portable.

◇ 例5

原文：肿瘤治疗用重离子加速器、质子加速器和 PET 配套质子回旋加速器为辐射科学和技术在医疗方面的应用提供了先进装备。

原译：Heavy ion accelerators used for tumor treatment, proton accelerators and cyclotrons supporting PET provide advanced devices for the application of radiation science and technology in medicine.

解析：根据专业知识和上下文可以判断"肿瘤治疗用"是其后三个名词的共同定语；"重离子加速器、质子加速器和 PET 配套质子回旋加速器"本身即为先进设备，而非提供设备的部分，因此"提供"不必翻译出来。

译文：Heavy ion accelerators, proton accelerators and PET cyclotrons for tumor treatment are advanced devices for the application of radiation technology in medicine.

练习
practice

1. 简答题。

(1) 除文中所提到的平台或网站，你还了解哪些获取科技知识的渠道？

(2) 英国著名博物学家、生物学家、教育家托马斯·亨利·赫胥黎有句名言：Try to learn something about everything and everything about something. (尽可能广泛地涉猎各门学问，并且尽可能深入地择一钻研。)请以译者的视角评价这句话。

2. 翻译下列句子，注意运用专业知识。

(1) 然而，卫星姿态系统受到诸多不确定因素的影响，如环境随机干扰力矩、系统惯性矩阵的不确定性和柔性变形引起的干扰力矩。

(2) 125 MW 高通量工程试验堆和 60 MW 中国先进研究堆(CARR)生产出多种医用和药用放射性同位素，并促进了中子成像技术及应用的开发。

(3) 密钥宽度、结果宽度和表内的条目数量可以决定 FPGA 所用片上逻辑资源和存储器(SRAM/DRAM)的数量。

(4) 随着基于策略的网络的演进发展以及软件定义网络(SDN)与基于意图的网络(IBN)的涌现，具有不同吞吐量与特点的防火墙逐步部署到企业网络的众多不同位置。

(5) 来自多个应用、访问企业网络与数据中心网络不同节点所产生的流量需要针对具体流量类型分类。

第二节　英语能力不足

英语能力涵盖口头理解、书面理解、口头表达、书面表达等方面，其中书面表达能力和译文质量紧密相关。根据《中国英语能力等级量表》，英语书面表达能力分为三个阶段：基础阶段、提高阶段、熟练阶段，涵盖四个维度：准确性、丰富性、连贯性、得体性(潘鸣威，2022)。专业译者应该达到熟练阶段，即准确的修辞手法，精准的措辞(准确性)；灵活丰富的措辞，多样的修辞手法和论证手法(丰富性)；综合衔接手段，逻辑缜密、行文流畅(连贯性)；得体的文化内涵表达，特定语体语境的表达变体(得体性)。若译者书面表达能力不足，则译文质量难以保证。

对于专业译者来说，提高英语能力可以采用短期计划和长期计划相结合的方法。短期内提高英语能力的一个有效途径是熟练掌握英语语法规则，提高语言的准确性。学习者可以以写作和翻译为出发点，重新学习英语语法规则，重视规则的应用，不能仅满足于死记硬背语法知识。市面上语法教材种类繁多，学习者可选择篇幅适中、强调语言结构的教材，如我国科技英语专家秦荻辉教授编著的《科技英语语法》《科技英语写作》等。长期计划包括增加语言输入和翻译实践、参加专业训练、学习专业书籍等。增加语言输入可以精读难度较高的非文学类文本，关注语言现象、汉英差别等。适量翻译实践既可以有效检验英语习得的效果，也可以提高英语能力。若有必要，可以参加专业训练，系统学习英语及翻译相关知识，提升英语能力。专业书籍包括学习策略、英汉对比、写作、翻译等方面的书籍。

【经典赏析】

原文：本周日，位于南海珠江口盆地的中国海油恩平 15-1 平台正式开启二氧化碳回注井钻井作业。这是我国第一口海上二氧化碳回注井，填补了我国海上二氧化碳封存技术的空白。

译文：China's first undersea Carbon Ddioxide (CO_2) re-injection well, developed by the China National Offshore Oil Corporation (CNOOC), started drilling in the Pearl River mouth basin, the South China Sea, on Sunday, filling in a void in the country's offshore CO_2 storage technology. The drilling work was undertaken by CNOOC's Enping 15-1 oil platform.

赏析：译文无任何语法或用词错误，准确度高。对"海上"的翻译灵活多变，分别翻译为 undersea 和 offshore，彰显了语言的丰富性。译文对原文语序和结构进行调整，将重要信息前置，将汉语的流水句处理为符合英文习惯的句式，并将重复信息"二氧化碳回注"进行整合，使译文逻辑缜密、行文流畅、语体正式，符合科技文体特征。

【案例解析】

◇ 例 1

原文：采用 P4 可以提高现有可编程 FPGA 架构的灵活性，因为它可以轻松实现数据包解析、数据包编辑以及流量表条目的修改。

原译：If P4 is used, the existed programmable FPGA architecture will be more flexible, because it is easy to realize packet analysis, packet editing, and change of flow list entries.

解析：原译中有两处重大语法错误：existed 和 it is easy to…。exist 为不及物动词，不能用其过去分词作定语。it is easy to…中的 it 若理解为形式主语，则"它"未被翻译出来；若理解为"它"，则用法错误。词汇使用不够精准，如 realize，change，list。句子结构松散，长短不均衡，语体不够正式。

译文：The use of P4 adds additional flexibility to the already programmable FPGA architecture because it allows easy implementation of packet parsing, packet editing, and modifications of flow table entries.

◇ 例 2

原文：元宇宙是一个流行概念，泛指整合虚拟现实和增强现实等技术形成的沉浸式虚拟世界。尽管元宇宙还处于起步阶段，但全球企业一直在大力开发和投资元宇宙相关技术。

原译：The metaverse is a popular concept, it extensively refers to an immersive virtual world that combines technologies like virtual reality and augmented reality. Though still nascent, global companies have been working to develop and invest in metaverse-related technologies.

解析：原译中有两处语法错误：前两个分句之间无连接词；still nascent 的逻辑主语和句子主语不一致，成为悬垂结构。用词也有不准确之处，popular

虽翻译为"流行的",但隐含意思是"大众化的",而"元宇宙"这一概念显然并非大众所熟悉的,而是引领了一种趋势,译为 trending 更准确；extensively 意思为"广泛的"而非"宽泛的"；global 意思为"全球的",但 global company 为"跨国公司"。

译文：The metaverse is a trending concept that refers, in general, to an immersive virtual world combining technologies like virtual reality and augmented reality. Though it is still in its infancy, many enterprises in the world have been gearing up efforts to develop and invest in metaverse-related technologies.

◇ 例3

原文：企业防火墙通常同时终止大量 TCP 连接,这将消耗大量的 CPU 周期与存储器。为了实现应用级安全处理,可能需要采用拥有大量核心的昂贵的高端 CPU 来终止众多 TCP/UDP 连接。

原译：Enterprise firewalls usually terminate a number of TCP connections at the same time, this consumes a number of CPU cycles and memory. To implement application-level security processing, we may need to use expensive high-end CPUs with a large number of cores to terminate the large number of TCP/UDP connections.

解析：原译前两个分句之间没有连接词,为典型流水句。a/the number of 反复出现,语言不够丰富。at the same time 和 we may need to use 不够简洁,语体不够正式。

译文：Enterprise firewalls usually terminate millions of TCP connections simultaneously, which consumes lots of CPU cycles and memory. To implement application-level security processing, expensive high-end CPUs with a large number of cores might be required to terminate the millions of TCP/UDP connections.

◇ 例4

原文：在高端安全设备中,需要采用 ML 模型处理海量实时数据,以便预测异常,因此采用加速器实现 ML 模型将给高吞吐量与低时延恶意软件预测带来巨大优势。

原译：In high-end security appliances, ML model is needed to be adopted to process huge amounts of real-time data to predict anomalies, therefore to use the accelerators to implement ML models will bring huge advantages to high throughput and low latency malware prediction.

解析：ML model is needed to be adopted to 有两处错误：ML model 应改为 the ML model 或 ML models；is needed to be adopted 谓语动词和补足语双重被动不符合英语习惯。两个分句之间无连词，成为流水句，不够规范。to use the accelerators to 虽语法正确，但不定式堆砌，语言不够流畅。bring huge advantages to 不够简洁。

译文：In high-end security appliances, the ML model is required to process huge amounts of real-time data to predict anomalies, so the use of the accelerators to implement ML models will greatly benefit high throughput and low latency malware prediction.

◇ 例 5

原文：众多现有的非对称算法容易受到量子计算机的破坏。对量子计算安全加密算法的研究和实现已经起步，而已经有学术论文介绍了如何采用 FPGA 实现此类算法。

原译：Many of currently known asymmetric algorithms are broken by quantum computer easily. The study and implementation of quantum computing safe cryptographic algorithms has already started, and academic papers have described how to use FPGAs for implementing such algorithms.

解析：原译第一句 Many of 后缺少定冠词 the。quantum computer 为单数可数名词，单独出现通常为不规范用法。"容易受到量子计算机的破坏"是一种可能性或趋势，而非事实。原译第二句话的第二个分句以 academic papers 作主语，使得核心信息未能凸显出来，how to use FPGAs for implementing… 中的 for implementing 不符合英语表达习惯(用 to implement 作状语更常见)，how to use 表达和前文风格不一致，略有违和感，且不够简洁。

译文：Many of the currently known asymmetric algorithms are susceptible to being broken by quantum computers. The study and implementation of quantum computing safe cryptographic algorithms has already started, and the use of FPGAs to implement such algorithms has already been described in academic papers.

练习
practice

1. 找出并修改下列学术论文中的错误。

Micro-Expression Recognition Based on MAML Meta-Learning Algorithm

Micro-expressions is an accurate and effective behavioral clue. It can reveal true changes in people's emotion. It is widely used in education, interrogation, clinical diagnosis etc. Previous researches on micro-expression recognition focused on traditional image recognition and deep-learning methods, both of them rely on large amounts of training samples and have lower accuracy. Therefore we introduce a meta-learning training method to achieve high accuracy with small samples. A three-dimensional convolutional neural network is used as the representation of the learning input of the feature extractor. In the meta-training stage, the optimal parameter $\theta*$ is obtained through double-loop iterative updating. We have conducted extensive experiments on two typical micro-expression datasets: CASME II and SMIC-HS. The results show recognition accuracy based on meta-learning can reach 92.97%, which is significantly better than traditional algorithms. In addition, the model can adapt to new micro-expression recognition tasks flexibly and quickly with small samples.

The research of micro-expressions started late, and the early research mostly uses traditional machine learning methods, mainly to obtain subtle facial motion information by hand. Although this method has a certain validity in recent researches, it consumes a lot of manpower and relies too much on the existing experience of researchers. In recent years, with the rapid development of the computer vision, researchers have applied deep learning and other related technologies to the field of micro-expressions recognition, thereby improving the recognition accuracy. At present, the collection of micro-expressions sample data has problems such as short duration, small motion range, and large impact by environmental factors, which makes it difficult to collect data with a small sample size and difficult to conduct better deep learning training. Therefore, how to extract the effective features of micro-expression from tiny facial movements has become the main research content of researchers. Secondly, the collection of micro-expression data requires videos with specific emotions to induce the subjects to express and maintain specific types of expressions. Professionals analyze and observe the time when the micro-expressions are generated and the subjects' true emotions to obtain sample data. Micro-expressions data sampling process is complicated and cumbersome and requires high environmental requirements, which causes great difficulties for the large-scale sampling of data sets. Therefore, the massive collection of micro-expressions data is also one of the problems that needs to be solved. Finally, how to use powerful feature extraction capabilities of neural networks while alleviating the network's dependence on the amount of data is a problem to be solved in the study of micro-expressions recognition using deep learning methods.

2. 翻译下列段落，注意语言的准确性、丰富性、连贯性、得体性。

(1) 雷达系统所用的微波器件具有结构复杂、精度高、成本高等特点。本文介绍了一种复杂微波器件的精密制造(precision manufacturing)技术，提出并讨论了其制造过程要点，重点分析了造成其加工变形(machining deformation)的主要因素以及在变形控制方面采取的具体措施。实践证明，本文所设计的制造技术很成功，对解决类似问题提供了有益的参考。

(2) 安全聚合(secure aggregation)是一种加密协议，可以安全地计算出其输入的聚合。它对于在联邦学习(federated learning)中保持模型更新的隐私性至关重要。本研究发现恶意服务器可以轻松逃避安全聚合，就好像安全聚合不存在一样。我们设计了两种不同的攻击，它们能够推断出关于个人训练数据集的信息。无论使用何种安全聚合协议，该攻击都是通用的，并且同样有效。我们的研究表明，当前使用安全聚合进行联邦学习的方式只提供了一种"假的安全感"。

(3) 本文介绍了一种稠密 SLAM 系统 Co-Fusion，它以 RGB-D 图像的实时流作为输入，并使用运动或语义线索将场景分割为不同对象，同时实时跟踪和重建其 3D 形状。我们使用多模型拟合(multiple model fitting)方法，其中每个对象可独立于背景进行移动，但仍然能够被有效跟踪，并且其形状也能够通过使用与该对象标签相关联的像素信息得以融合。以前人们在处理动态场景时，通常尝试将移动区域视为异常值(outlier)，因此不会对其形状进行建模，也不会跟踪它们随时间的运动情况。相比之下，我们让机器人维护每个分割对象的 3D 模型，并随着时间的推移通过融合对其进行改进。

(4) 华为作为全球领先的电信服务提供商，一直致力于为用户提供稳定的网络运行和不间断的服务。华为的人工智能能力和数字化平台可确保一流的数字交付和维护，帮助运营商提高服务质量，从而使其与客户保持更好的互动。

华为人工智能客户服务、在线医疗及新型营销为运营商带来完全不同的体验。今年，华为在全球范围内总共交付了数十万个 5G、4G 站，这都依托于华为集成服务交付平台(ISDP)。ISDP 已经开发了 100 多个应用和 500 多个微服务，覆盖了交付全流程，从 PO 的获取到勘测、部署、验收、发票及回款所有环节，最大限度地减少了密切接触和因此产生的公共卫生风险。

(5) 信息推动了广播电视的兴起。在 19 世纪 80 年代，海因里希·赫兹(Heinrich Hertz)证明了无线电波是电磁辐射的一种形式(如麦克斯韦理论所预测的那样)。在 19 世纪 90 年代，印度物理学家贾加迪什·钱德拉·玻色(Jagadish Chandra Bose)进行了一项实验，利用微波点燃火药并敲响铃铛，证明了电磁辐射可无线传播。这些科学见解为现代电信奠定了基础。

计算技术也发生了类似的演变。早期的大型计算机非常庞大，占据了房间的大部分空间。它们价格昂贵，而且散热很差，需要冷却，似乎只有政府或财力雄厚的大型企业才用得起。在 20 世纪 80 年代，个人计算机极大地改变了这种状况。突然之间，所有企业和多数个人可以购买和使用计算机，不仅将它们用于密集计算，还用于管理信息。

第三节 母语负迁移

美国语言学家拉里·塞林格(Larry Selinker)(1972)认为第二语言学习者会构建一个既依赖母语却又不同于母语也不同于目标语的独特的语言系统，它介于母语和目标语之间，即"中介语"(interlanguage)。在中介语的形成过程中，母语的迁移是一个重要因素。母语对目标语的影响既有正面的也有负面的，能够促进目标语学习的正面影响被称为"正迁移"，干扰和阻碍目标语学习的负面影响被称为"负迁移"。

翻译作为一种特定的双语活动是第二语言习得的必然阶段，译者的翻译活动都构建在其对母语和目标语的认知之上，翻译过程必然受到母语的影响。可以说，译文本身就是一种中介语，而母语迁移是中介语的主要根源。母语迁移表现在译文里就是与原文形式上的一种对等。正迁移产生的对等接近于目标语习惯，易于被目标语读者接受，是合格的译文；负迁移则会产生过度对等，使译文生硬晦涩，可读性差。

翻译中的母语负迁移现象与译者双语能力、对两种文化的掌握程度、对翻译的态度等因素有关。通过提高双语能力、文化差异和语言差异感知度、译者责任感等可以有效避免母语负迁移。此外，译者应对中介语现象进行有意识的归纳分析，自主干预，或阅读相关书籍，最大程度地避免母语负迁移。琼·平卡姆(Joan Pinkham)女士所著《中式英语之鉴》(*The Translator's Guide to Chinglish*, 2000)一书从词汇和句法层面列举了中式英语中不为英语本族语所接受的表达方式，可以看作是对母语负迁移较为系统、全面的总结。

【经典赏析】

原文：把绿色科技推广应用与自主研发紧密结合起来，把整合现有科技资源与布局未来科技创新紧密结合起来，把加强生态环境治理修复与促进产业结构调整紧密结合起来，突破制约发展方式转变和可持续发展的重大科学问题和关键技术，为生态文明建设提供知识基础、科学依据和技术支撑，促进人类社会可持续发展，促进人与自然和谐共生。

译文：Application of green technology should go hand in hand with independent research and development, integration of existing technological resources should have future innovation in mind, and improvement and restoration of the eco-environment should be closely connected with industrial structure adjustment, so as to solve significant scientific issues and key technologies that hinder the transformation of the growth model and the sustainable development, provide knowledge, scientific theory and technological support for an eco-civilization, and promote sustainability of humanity and harmony between man and nature.

赏析：原文中"紧密结合起来"出现三次，但译者灵活使用不同表达，有效避免了重复。将"突破"翻译为 solve 而非与其对等的英文表达 break through，将"知识基础"仅翻译为 knowledge，简洁明了。原文为流水句，译文中用 so as to、for 等词将逻辑显性化，使句子结构更符合英语表达习惯。无论是在词汇层面还是句法层面，甚至是信息加工层面，均避免了母语负迁移。

【案例解析】

◇ 例 1

原文：在神经网络中，矩阵乘法和卷积属于最耗费内存和计算资源的运算之一。

原译：In neural networks, matrix multiplication and convolution belong to the operations that consume the most memory and compute resources.

解析：原文中的"属于"看似对应英语中的 belong to，但在该语境中表示类别而非所属，可理解为"是……之一"。"最耗费内存和计算资源"可根据专业知识灵活翻译。

译文：In neural networks, matrix multiplication and convolution are among the most memory-consuming and computing resource-consuming operations.

◇ 例2

原文：本次活动时间为全年，试点地区将举办不同形式的线上线下活动，加快节能低碳的绿色建材推广应用。

原译：The activity will last for a whole year, different online and offline activities will be held in pilot areas, and the popularization and application of low-carbon energy-saving building materials will be accelerated.

解析：受原文影响，原译中 activity 重复出现，词汇不够丰富；low-carbon 和 energy-saving 意思重复，译文不够简洁；"绿色建材推广应用"翻译略显拗口，popularization 较难理解。

译文：The yearlong campaign will see a string of online and offline activities organized in pilot areas to accelerate low-carbon development in the building materials sector.

◇ 例3

原文：在网络运营和运维方面，华为有赖其基于人工智能的自动化智能(AUTIN)系统，可提前对网络故障进行准确预测和检测。通过新的告警处理方法，该系统可大大减少工程师上站解决问题的次数，从而降低维护工程师接触不安全环境的风险。目前，AUTIN 系统支持 130 多个国家的 100 多个项目。

原译：In the aspect of network operation and maintenance, Huawei depends on its AI-based AUTIN system to perform accurate prediction and detection of network faults in advance. By the new alarm handling methods, the system can greatly reduce the number of times that engineers go to sites to solve problems, thereby reducing the risk of maintenance engineers touching unsafe environment. Currently, the system supports more than 100 projects in over 130 countries.

解析："在网络运营和运维方面"在该语境中的意思为"就网络运营和运维而言"原译中译为 in the aspect of 不够准确。"华为有赖其基于人工智能的自动化智能(AUTIN)系统"强调的是系统而非华为，原译以 Huawei 作主语，重点信心不够突出。"对网络故障进行准确预测和检测"中的"进行"没有必要翻译出来。"工程师上站解决问题的次数"翻译不够简洁。"接触"翻译不准确。

译文：In terms of network operation and maintenance, the Huawei AI-based AUTIN system can predict and detect network faults accurately in advance. By the new alarm handling methods, the system can greatly reduce the number of site visits needed, thereby reducing the risk of maintenance engineers being exposed to a potentially unsafe environment. Currently, the system supports more than 100 projects in over 130 countries.

◇ 例4

原文：FPGA 是新一代防火墙内联安全处理的理想选择，这是因为采用 FPGA 可以成功满足对更高性能、灵活性和低时延操作的需求。此外，FPGA 还可以实现应用级安全功能，从而进一步节省计算资源并提高性能。

原译：FPGA is an ideal option for inline security processing in next-generation firewalls. This is because using FPGA can successfully meet the requirements for higher performance, flexibility, and low latency operations. Besides, FPGA can also implement application-level security functions to further save computing resources and increase performance.

解析：FPGA 是英文 Field Programmable Gate Array 的缩写，即"现场可编程门阵列"，为可数名词，通常需要和冠词连用或用复数形式。"理想选择"和"这是因为"原译为完全对等的表达，不够简洁。"对更高性能、灵活性和低时延操作的需求"中的"更高性能、灵活性和低时延操作"并非表示对象，而是"需求"的具体内容，即"需求"的同位语，应用介词 of 表达。"此外"和"还"在汉语中为固定搭配，但不必将两个词都翻译出来。

译文：FPGAs are ideal for inline security processing in next-generation firewalls because the requirements of higher performance, flexibility, and low latency operations are successfully met using FPGAs. FPGAs can also implement application-level security functions to further save computing resources and improve performance.

◇ 例5

原文：零信任体系结构在 2010 年由时任 Forrester Research 主要分析师的 John

Kindervag 开发，是一个广泛的框架，承诺有效保护企业最有价值的资产。其工作原理是假设每一个连接和端点都被视为威胁。

原译：The zero trust architecture was developed by John Kindervag in 2010 while he was a principal analyst at Forrester Research, it is a broad framework, and it promises to effectively protect an organization's most valuable assets. Its work principle is to assume that every connection and endpoint is considered a threat.

解析：译者受原文影响，未能对句子结构进行调整，以流水句的形式呈现信息，it 重复出现，结构松散。主语"工作原理"翻译为 Its work principle 不够准确，working principle 更为常见。

译文：Developed by John Kindervag in 2010 while he was a principal analyst at Forrester Research, the zero trust architecture is a broad framework that promises effective protection of an organization's most valuable assets. It works by assuming that every connection and endpoint is considered a threat.

练习
practice

1. 赏析下列翻译，指出译文的优点。

(1) 低碳发展是一种以低耗能、低污染、低排放为特征的可持续发展模式，对经济和社会的可持续发展具有重要意义。

Low-carbon development is a model of sustainable growth featuring low energy consumption, low pollution and low emission.

(2) 本周五，中国互联网络信息中心发布《中国互联网络发展状况统计报告》。报告指出，截至 2019 年 6 月，我国网民规模达 8.54 亿，互联网普及率达 61.2%；我国手机网民规模达 8.47 亿，网民使用手机上网的比例达 99.1%。

The number of Internet users in China had hit 854 million as of June 2019, with the Internet availability rate reaching 61.2 percent, according to a report on China's Internet development released Friday. A total of 847 million Chinese people used mobile phones to surf the Internet, accounting for 99.1 percent of the total netizens, stated the report issued by the China Internet Network Information Center.

(3) 作为世界上首颗量子科学实验卫星，"墨子号" 8 月 16 日成功发射标志着中国在量子通信领域已经走在了世界前列。

China became a world leader in the area of quantum communications when it launched Micius, the world's first quantum experiment satellite, on Aug 16.

(4) 该省明确了十大战略性新兴产业，例如新一代信息技术、新能源汽车和智能网联汽车、人工智能和新材料等。

The province listed 10 main sectors as strategic emerging industries, among them new-generation information technology, new-energy and intelligent and connected vehicles, artificial intelligence and new materials.

(5) 用户无须再记忆越来越冗长的密码，而可以使用身体特征或者已有设备认证其登录信息，通过蓝牙、USB 接口或近场通信技术直接完成在线身份认证。

Instead of having to remember an increasingly long string of characters, users can authenticate their login with their body or something they have in their possession, communicating directly with the website via Bluetooth, USB or NFC.

2. 翻译下列句子，注意避免母语负迁移。

(1) 中国著名的量子物理学家潘建伟从 2001 年开始在中国科大组建科研团队，并于 2009 年倡议，由学校成立一个公司，助力团队的研究，推动学术成果的商业转化。

(2) 在模拟特别复杂的场景时，量子计算机可以采用计算捷径，而传统计算机必须一步步地找解决方案，在过程中花费更多时间。

(3) 随着中国继续朝着更可持续的能源结构发展，可再生能源所占比例日益增大，虚拟电厂的建设正在中国兴起。

(4) 随着科技公司争相推出类似 ChatGPT 的人工智能聊天机器人，这对大型语言模型训练和操作过程中的算力提出了更高的要求，中国的算力行业预计将飞速发展。

(5) 然而，随着发电的重点逐渐转向光伏和风电等可再生能源，问题出现了，这些供电方式是间歇性的，并不是在需要时总是可用，这使得管理电力供应极其困难。

第四节　职业精神不够

译文质量不仅取决于译者的双语能力，还与译者的态度紧密相关。若不能对翻译事业抱有敬畏之心、没有为客户负责的担当、没有精益求精的愿望，译文将如同无勒之马、无舵之舟，没有方向，质量飘忽不定。唯有恪守职业道德规范，外加良好的语言能力，译文质量才能得以保证。

【经典赏析】

原文：中国科学家近日宣布，中国高海拔宇宙线观测站(LHAASO，拉索)在银河系内发现大量超高能宇宙加速器，这一发现可能颠覆人类对银河系的认识，并有助于揭示宇宙线起源这个困扰了科学家一个世纪的问题。

译文：Chinese scientists have detected a dozen ultra-high-energy (UHE) cosmic accelerators within the Milky Way, a find that could overturn humanity's understanding of the galaxy and help reveal the origin of cosmic rays, which have puzzled scientists for a century. The findings were based on the observations of China's Large High Altitude Air Shower Observatory (LHAASO).

赏析：译者以专业的精神对原文句子结构进行调整、拆分，将流水句处理为符合英语习惯的句式，并使句子长短适中，重点突出，增强了译文的可读性。同时，译者以专业素养对"大量超高能宇宙加速器"中的限定词"大量"处理为 a dozen，而非直接翻译为 a huge amount of，使译文措辞更客观、更专业。

【案例解析】

◇ 例1

原文：图7显示了用于 MACSec 和 IPSec 的数据包编辑/修改操作。在通过出站端口发送数据包之前，可能需要对包头字段进行多次修改，包括校验和与 CRC 的计算与插入。

原译：Figure 7 showed the packet editing operations for MACSec and IPSec. It might be necessary to modify the packet header fields for many times before sending them out at the outbound port, including verification and calculations and insertion of CRC.

解析：原译无重大语法错误，可见译者双语能力尚可，但有诸多细节错误，可见译者之态度。例如，原译有两处语法错误，showed 为时态错误，them 指代不清。including verification and calculations and insertion of CRC 和上文逻辑关系不清晰。It might be necessary to modify the packet header fields for many times 不够简洁，语体不够正式。更为严重的是，译者在处理"校验和与 CRC"时，无视"和""与"叠加使用的特殊现象，擅自删掉其中一个字，将原文理解为"校验与 CRC"。有责任心的译者只要稍作查证或思考即可判断出"校验和"为专业术语。

译文：Figure 7 shows the packet editing/modifying operations for MACSec and IPSec. Many modifications might be needed in the packet header fields, including calculations and insertion of checksum and CRC, before sending the packet out at the outbound port.

◇ 例 2

原文：这些面板来自符合 ASTM C1396 规范的防潮液综合生产线。

原译：These panels come from a comprehensive product line of against tide solutions that meet ASTM C1396 specifications.

解析：原译将"防潮"译为 against tide 就如将"雄安"译为 male safety 一样荒诞，体现出译者责任感的缺失。

译文：These panels come from a comprehensive product line of moisture-resistant solutions that meet ASTM C1396 specifications.

◇ 例 3

原文：阿里云敏感数据保护文档及使用对象

原译：Ali Cloud Sensitive Data Protection Document and the Use of the Object

解析：原译为 DeepL 的翻译结果，有道词典对"使用对象"的翻译有 using object、object of use 等。无论是 DeepL 还是有道，都只是翻译的辅助工具，若非客户允许，不应向其交付机器翻译结果，此为职业道德。在翻译中可适当使用电子工具，但译者应有自己的思考、判断，不可过度依赖电子工具。

译文：Ali Cloud Sensitive Data Protection Document and Its Target Audience

◇ 例 4

原文：在传输方面，提出了新型 ldpc 编码方法和峰平比低的调制方法，实现了比特率为 12 Mb/s，误码率为 7.9×10^{-7} 的高效传输系统；

原译：In terms of transmission, we proposed a new ldpc coding method and a modulation method with a low peak to average power ratio, realizing an efficient transmission system with a bit rate of 12Mb/s and an error rate of 7.9×10^{-7}.

解析：原文中的"7.9×10^{-7}"看起来不像常规数字，恐有错误。译者应妥善处理，或查证或向客户求证，不可无视错误，将错就错；peak to average 改为 peak-to-average 可增加可读性。

译文：In terms of transmission, we proposed a new ldpc coding method and a modulation method with a low peak-to-average power ratio, realizing an efficient transmission system with a bit rate of 12Mb/s and an error rate of 7.9×10^{-7}.

练习
practice

1. 赏析下列翻译，指出译者精神的体现。

(1) 世界互联网大会发布的蓝皮书指出，根据最新发布的世界互联网发展指数，中国在全球的排名仅次于美国，且发展形势处于稳步上升状态，排名前五的其他三个国家分别是德国、英国和新加坡。

China has been ranked second in the latest Global Internet Development Index, after the United States, and has steadily maintained its upward trajectory, with Germany, the United Kingdom and Singapore in the top five, according to the Blue Book for the World Internet Conference.

(2) 自 2022 年 11 月 30 日顺利进驻空间站组合体以来，神舟十五号航天员乘组已在轨工作生活 70 天，先后完成了与神舟十四号航天员乘组在轨轮换、科学实验机柜测试、飞船设备巡检等工作。

The three-man crew have been living in orbit for 70 days since they entered the space station combination on Nov. 30, 2022. They completed various tasks, including in-orbit crew rotation with the Shenzhou-14 taikonauts, tests of scientific experiment cabinets, and spacecraft equipment inspections.

(3) 工业和信息化部已向中国商飞公司发放首个企业 5G 专网频率许可，

是 5925～6125 MHz 和 24.750～25.15 GHz 的工业无线专用频段，具有高速率、低时延等技术优势。

The Ministry of Industry and Information Technology, the country's top industry regulator, has granted China's first corporate 5G private network license to Commercial Aircraft Corp of China to use industrial wireless dedicated frequency bands 5925−6125 megahertz and 24.750−25.15 gigahertz, which feature high speed and low latency.

(4) "月球漫步者"由美国卡耐基梅隆大学下属的 Shift Robotics 公司设计，每双重量为 1.9 kg 的鞋中包含一个 300 W 的无刷电机，为 8 个聚氨酯轮子提供动力。

Designed by Shift Robotics, an offshoot of Carnegie Mellon University, Moonwalkers consist of a 300W brushless motor in each 4.2-lb (1.9-kg) shoe that powers eight polyurethane wheels.

(5) 待机能耗是指电器关机或处于待机模式时所耗的电。

Standby power refers to the electric power consumed by electronic and electrical appliances while they are switched off or in standby mode.

(6) 特斯拉首席执行官马斯克近日到访中国，并在本月初参观特斯拉上海超级工厂。他表示，上海超级工厂不仅是世界上生产效率最高的工厂之一，造出的车也是全世界品质最好的车之一。

Tesla CEO Elon Musk paid a visit to the Tesla Shanghai Gigafactory earlier this month during his trip to China, and lauded the plant for its production efficiency and quality.

(7) 该算法的目标是训练一个能够快速适应新任务的模型，其基本思想是找到一个更好的初始参数，使模型能够以更少的梯度步骤快速学习新任务。

The algorithm aims to train a model that can quickly adapt to new tasks, and its basic idea is to find a better initial parameter so that the model can quickly learn new tasks with fewer gradient steps.

(8) 国内首家国家级氢能动力质量监督检验中心 1 月 9 日在重庆建成投用。该中心由中国汽车工程研究院股份有限公司投资建设，致力于推动中国氢燃料电池汽车产业发展。

National Hydrogen Power Quality Inspection and Testing Center, an organization dedicated to boosting the development of China's hydrogen fuel-cell vehicle industry, started operation on Jan. 9 in Chongqing Municipality. The center, the first of its kind in China, was established by China Automotive Engineering Research Institute Co., Ltd.

2. 指出下列翻译中有失译者职业精神的体现。

(1) 综合业务模件电路图

Integrated Service Model Circuit Diagram

(2) 与燃油车相比，电动车在运行过程中几乎不排放二氧化碳，但生产电池会产生大量污染，因此当车辆行驶里程达到数万英里时，电动车的全周期碳排放与燃油车相当。

Battery electric vehicles emit virtually no CO2 during operation compared with combustion-engine vehicles, but battery cell production can create so much pollution that it can take EVs tens of thousands of miles to achieve "carbon parity" with comparable fossil-fueled models.

(3) 我国将巩固其在新能源汽车等行业的领先地位，并采取措施确保新能源汽车行业蓬勃发展。

Our country will strengthen its leading status in industries like new energy vehicles, and take steps to ensure the NEV industry flourishes and develops.

(4) 随着中国加大力度确保能源安全，同时促进绿色能源转型，2022 年，全国风电、光伏发电新增装机达到 1.25 亿千瓦，再创历史新高，使总装机容量超过 12 亿千瓦。

As China stepped up efforts to ensure energy security while facilitating green energy transition, its newly installed combined wind and solar power capacity reached 125 million kilowatts in 2022, which was a new high, bringing the total installed capacity to over 1.2 billion kW.

(5) 研究表明，通过设置前置机，既保证了系统的安全，又提供了数据交换质量。

The study shows that setting up a front-end both ensures the security of the system and provides the quality of data exchange.

(6) 从这条曲线可以看出产品失效规律可分为三个阶段，即早期失效期、偶然失效期和损耗失效期。

According to this curve, it can be seen that the law of product failure can be divided into three stages: early expiration date, occasionally expiration date and loss expiration date.

(7) 近日，应 *Information Fusion* 主编 Salvador Garcia 教授邀请，我校机电工程学院张静教授团队和英国 Aberystwyth University 计算机科学学院终身教授韩明课题组联合在多模态信息融合领域顶级期刊 *Information Fusion* (影响因子 17.564)，发表了题为"Deep Learning for Visible-Infrared Cross-modality Person Re-Identification: A comprehensive Review"综述论文。

Recently, invited by Professor Salvador Garcia, Editor in Chief of *Information Fusion*, Professor Zhang Jing's team, School of Mechano-Electronic Engineering at our university and Professor Han Ming's team, Computer Science Department at Aberystwyth University, published a review paper 'Deep Learning for Visible-Infrared Cross-modality Person Re-Identification: A comprehensive Review' in *Information Fusion* (IF:17.564), which is the No.1 journal in the field of information fusion.

(8) 一些全球规模最大的资料库都依托于我们的平台，因此平台能够将存储挑战转变为业务优势。平台能够在降低成本的同时，以可靠的方式支持传统工作负载和新兴云原生工作负载，助力企业的移动、社交、分析和认知计算。

Some of the world's largest repositories are supported by our platform, so our platform turns storage challenges into business advantage. It supports both traditional workloads and emerging cloud-born workloads for enterprise mobile, social, analytic and cognitive computing in a reliable way while reducing storage costs.

3. 以译者的专业精神翻译下文。

苹果公司推出了数字健康工具，以帮助人们减少沉迷于手机的时间。一款名为"屏幕时间"的新应用面向的是苹果手机和苹果平板电脑的用户，将会用数字仪表板的形式显示出用户使用哪一款应用消耗了多少时间，用户收到了多少个通知，以及他们拿起手机的频率，还有他们的使用模式和平均水平的对比。这款应用还让用户设置单个应用的每日使用时间上限，时间快到时会发出通知。父母将能够在自己的手机上访问孩子的活动报告，了解和管理孩子的浏览习惯。"屏幕时间"应用将会出现在苹果新一代手机操作系统 iOS 12 上，该操作系统将于今年晚些时候发布。

苹果公司还加倍重视保护用户隐私，引进一系列工具，以防止广告商在线跟踪苹果手机的用户。苹果还将打击一种名为"指纹识别"的技术，这种技术允许广告商基于每台电脑的独特设置，包括装置、操作系统和字体，来识别和跟踪个人网络用户。

附录　部分习题参考答案

第一章　翻 译 概 述

第一节　科技翻译与功能翻译理论

1. 莱斯的文本类型划分和我们平常所理解的文体类型有何异同？

莱斯的文本类型划分，是将文本的文体特征与文本意欲再现的功能相联，并将文本划分为信息型文本、表达型文本和操作型文本三种类型，每种类型又包括若干不同文本类别；而我们平常所理解的文本类型的划分，多依据文本的语言特征和所表达的信息内容，将文本分为科技文本、法律文本、旅游文本等内容。

2. 莱斯基于文本功能的文本类型划分对于翻译实践有哪些启示与指导意义？

莱斯基于文本功能的文本类型划分，对翻译实践而言，让我们对翻译活动有了更多维度的思考与评判。以往翻译活动多关注文本信息，而实用文本翻译，往往会因为原文文本和译语文本所承载的不同功能而在翻译过程中对原文信息进行调整，莱斯的文本功能理论为应用型文本翻译过程中的改写提供了可依据的标准和原则。

第二节　翻 译 目 的 论

1. 弗米尔提出的翻译目的论，主要有哪些特征？

弗米尔所提出的目的论，主要指翻译过程中应必须遵循的三大原则：目的原则(skopos rule)、连贯原则(coherence rule)和忠实原则(fidelity rule)。其最大特点，就是以"目的原则"为翻译过程中的总原则，并决定翻译过程中所有翻译策略与方法的选择。也就是说，翻译目的的重要性超过了文本层面的忠实。

2. 诺德的翻译目的论与汉斯·弗米尔的翻译目的论之间，有哪些异同？

诺德的翻译目的论与弗米尔的翻译目的论都关注所译文本的内容，只是诺德认为弗米尔的目的论过于强调译文功能的实现，在翻译过程中容易忽视原文语言特征在译文中的再现，因而她在弗米尔的理论基础之上做了进一步的完善。诺德在《翻译中的文本分析》(*Text Analysis in Translation*，1988)中勾画了一个更详细的功能模式，提出了两种基本的翻译方法：文献型翻译(documentary translation)和工具型翻译(instrumental translation)。她将文学文本的翻译也囊括在内，不但强调翻译过程中翻译功能的实现，还强调对文本结构、语言特征的分析与再现。

第三节　科技翻译的译者素养和普遍方法

1. 相比文学翻译，科技翻译在翻译标准上有哪些不同？

相对而言，文学文本翻译更注重文学审美在译文中的再现，科技文本翻译则更注重原文信息在译文中准确、清楚的表达。科技文本翻译所遵循的"忠实准确、通顺流畅、规范专业、简洁明晰"的翻译标准，就是对科技文本翻译不同于文学文本翻译最好的表述。

2. 总体而言，科技翻译译者需要具备哪些素养？

在科技文本翻译过程中，初学者应从一开始就应关注科技译者素养、科技翻译标准、科技翻译过程等众多内容，并且从词汇、句法和篇章三个维度关注科技文本的翻译方法与注意事项，力争从译者能力、译文评判和翻译过程三方面对科技翻译有全面了解。此外，鉴于计算机技术在科技翻译中的广泛应用，如何使用机器翻译来提升翻译水平，也成为提高科技翻译水平的重要内容。此外，提升科技翻译技能，还需要关注翻译过程，有严谨的翻译态度。

第二章　科技翻译伦理

第一节　中国翻译伦理观

1. 简答题。

略。

2. 用现代文解释彦琮法师所著《辩正论》中的"八备"，说明其核心思想，指出哪几条与译者的道德品质有关。

(1) "诚心爱法，志愿益人，不惮久时，其备一也。"即要求译经人诚心诚意，接受佛法观点，立志做有益于他人的事业，不怕花费长久的时间。

(2) "将践觉场，先牢戒足，不染讥恶，其备二也。"要求译经人品行端正，忠实可信，不惹别人讥笑厌恶。

(3) "筌晓三藏，义贯两乘，不苦暗滞，其备三也。"要求译经人对佛教经典有渊博知识，通达大小乘经论的义旨，不存在含糊疑难的问题。

(4) "旁涉坟史，工缀典词，不过鲁拙，其备四也。"要求译经人通晓中国经史，具有高深的文学修养，文字表达准确，不疏拙。

(5) "襟抱平恕，器量虚融，不好专执，其备五也。"要求译人心脑宽和，虚心好学，不固执己见，不武断专横。

(6) "耽于道术，淡于名利，不欲高衔，其备六也。"要求译经人刻苦钻研学问，不贪图名利，不奢望高贵职衔。

(7) "要识梵言，乃闲正译，不坠彼学，其备七也。"要求译经人精通梵文，熟悉正确的翻译方法，不失梵文所载的义理，但又不能拘泥于梵本格式。

(8) "薄阅苍雅，粗谙篆隶，不昧此文，其备八也。"要求译经人对中国文字学具有一定的修养，熟悉文字的使用，保证译文通畅典雅，忠实准确。

彦琮所提出的"八备"，核心是要求翻译力求忠实，而要做到忠实，译者必须具有高尚的品德和一定的汉、梵文的修养和造诣。其中第一、五、六这三条，特别注重翻译人员的思想修养。

第二节　国外翻译伦理观

1. 简答题。

(1) 如何理解皮姆所提出的第三条翻译伦理"译者伦理不包括在两个文化间决策优劣"？

尽管有时需要作出选择，译者也应避免将不同语言、文化、社会和社会阶层对立起来。伦理准则都应该是文化间的，支持或反对一方都应该基于文化间合作的考虑。

(2) 简述切斯特曼的五个翻译伦理模式对你的启示。

前四种翻译伦理模式，译者不可能同时兼顾，而且四种模式之间也缺少兼容性，体现出矛盾性。这就需要译者以承诺伦理来协调各方的冲突。"再现伦理"可谓是翻译职业道德规范的根基。职业道德的规范与誓言对译者的首要预期与要求必然是要忠实于原作，当然并不是机械地拘泥于字句。"服务伦理"是翻译职业道德规范的必然追求。翻译作为社会职业，其产出成果必然拥有读者、观众、客户或终端用户，译者必然也要对他们负责。然而，现实情况中往往会出现诸多令译者进退两难的局面，此时就要遵守承诺伦理，凭借着强烈的社会责任感和高度的职业道德，尽善尽美，追求卓越。

(3) 在翻译中，面对委托人的不正当要求，你如何处理？

严格自律，不为经济利益而进行胡译、歪译，保持良知。

2. 阅读下面的翻译小故事，并从翻译伦理角度讨论 Stefan Moster 的翻译行为。

译者必须考虑译文读者的审美情趣、文化身份、价值观念等因素，以确定相应的翻译策略。有时译者还不得不对原作进行加工改造，甚至是曲解或杜撰，以使译作与读者的"期待视野"融合。这样，通过接近读者的阅读期待，使自己的译作能够在译语文化语境中得到认同或发挥特定的作用。

第三节 科技译者职业道德准则

阅读国际翻译家联盟欧洲区域中心发布的译者职业道德准则，找出它和中国翻译协会《译员职业道德准则与行为规范》的共核部分。

国际翻译家联盟欧洲区域中心发布的译者职业道德准则和中国翻译协会《译员职业道德准则与行为规范》均涵盖以下几方面：端正态度，胜任能力，忠实传译，保持中立，保守秘密，遵守契约，合作互助，妥用技术，提升自我。

第三章 科技译者素养

1. 译者的语言素养主要体现在哪几个方面？

译者的语言素养体现在源语应用能力、目标语应用能力和双语转换能力三个方面。

2. 译者如何提高自身科技素养？

译者应树立终身学习的意识，持续跟踪国际、国内科技发展，广泛了解和学习各科技领域的专业词汇和基本知识。在此基础上结合个人禀赋，重点深入掌握某些特定领域作为自己日后进行科技翻译的主攻方向。在科技翻译过程中要秉持严谨认真、科学准确的态度，具备锲而不舍的毅力，养成查阅文献和多次检查确认的习惯，精益求精，确保翻译质量。

3. 科技译者应该掌握哪些翻译理论知识？

首先应该掌握翻译活动本身即翻译本体的知识，主要包括翻译本质、翻译过程、翻译单位、翻译标准以及翻译的思维方法等；其次是关于翻译主体即译者的知识，主要包括译者在翻译中的主体地位，译者在翻译过程中的创造性和局限性，译者的知识结构，译者的能力构成以及译者的职责和道德素质，译者的常见错误分析等；最后是关于翻译客体也就是文本的意义或者说信息的知识，主要包括意义的定义、种类，文化差异，思维差异以及语言差异对翻译意义的影响，意义的可译性与不可译性，各种不同意义的特点以及它们与翻译的关系等。

4. 科技译者的信息素养主要包含哪几个方面？

科技译者的信息素养主要体现在计算机辅助翻译技能、信息检索能力、术语能力、译后编辑能力等方面。

第四章　科技翻译标准

1. 简答题。

(1) 文学翻译和非文学翻译各有哪些特点？

文学翻译涉及想象中的个人、自然、人类居住的星球，非文学翻译涉及知识、事实和思想、信息、现实；文学翻译强调的是价值和风格，非文学翻译强调的是事实和信息的清晰性。

(2) 如何理解科技翻译的标准？

科技翻译的标准可以概括为忠实准确、通顺流畅、规范专业、简洁明晰。

2. 根据科技翻译的标准，赏析下列翻译。

(1) 赏析：译文使用 former 和 latter 来代替前文出现的 refractor 和 reflector，既避免重复，又前后衔接，中文的"两种望远镜"在译文中直接用 both 替换，简洁明晰。译文中最后两句之间使用 this image 与前文的 real image 衔接，而中文"基本类似放大镜"作为"接目镜"的定语，在译文中借助被动语态将 eye-piece 后置，之后用 which 引出的非限制性定语从句表示，语义连贯顺畅。

(2) 赏析：译文选取"集成数字 PSTN 的重要优势在于节省了运营成本"为表意核心，将"与笨重的机电和半电子模拟交换相比，数字电子设备占地面积大大减少"作为"节省运营成本"的原因用 resulting from 引出，并采用名词化结构 reduction，使得译文结构灵活，简洁明晰，句意通顺流畅。此外，"机电和半电子模拟交换"等专业词汇的翻译也体现了规范专业。

(3) 赏析：译文采用名词化结构将因果关系的句子改为简单句表述，并将中文中括号内的"即经过视觉矫正的透镜"省译为 visually corrected 作为 lens

的补充定语，中文中"使用……进行天文摄影"也省去"进行"，转为 use…for astronomical photography 的简单表达，整体译文简洁明晰。

(4) 赏析：译文将中文的一句话分为两句进行翻译，其中第二句采用带有比较结构的简单句"No…can…as sharply as…"表现原文中"再怎么……也无法……"的逻辑关系，结尾部分使用"provided + 两个 that 引导的从句"表述前提条件，且"大型望远镜的物镜"直接译为 the larger object-glass，整体译文更加简洁，且"放大率""物镜"等专业词汇的翻译规范专业。

(5) 赏析：中文第一句中括号内的部分在译文中转为逗号隔开的 known as training data，译文以 build a model 为表意重心，将"无须编程"放到了句尾用介词短语 without being explicitly programmed to do so 来表示，其中的 to do so 与前文的 make predictions or decisions 照应，整句翻译通顺流畅。第二句中的"领域"前面的部分定语转为 where 引导的非限制性定语从句，表意更加明晰，其中 difficult 和 infeasible 分别与"难以"和"无法"对应，忠实于原文语义。

3. 赏析下述科技短文翻译中体现的翻译标准。

可根据忠实准确、通顺流畅、规范专业、简洁明晰等科技翻译标准分析科普文章的翻译特点。

第五章　科技翻译过程

第一节　理　解　原　文

1. 简答题。

(1) 翻译过程有哪些环节？

原文理解、表达。

(2) 理解原文应注意哪些部分？

正确选取词义；注重形合与意合的转化；考虑文化差异，运用翻译方法和技巧来使译文为读者所接受。

2. 赏析下列翻译，指出译文的优点。

略。

3. 在正确理解原文的基础上翻译下列句子。

(1) They are formed by a sub-set of a primary system's objects (its moons, rings, populations of gas, dust particles, plasmas and charged particles)

which are principally coupled to one specific planet and are confined by the effect of its gravity and magnetic fields to occupy a limited region of space around it.

(2) The multilayer perceptron is an artificial neural network structure and is a nonparametric estimator that can be used for classification and regression.

(3) Neptune-like planets represent a critical test for the core accretion scenario of giant planets formation.

(4) Little is known about the composition and internal dynamics of these moons.

第二节 查阅平行文本

1. 简答题。

(1) 平行文本是什么?

形式上非常一致的译文及原文、形式上不完全一致但功能对等的译文及原文、在同样交际情境中独立产生的两种不同语言的文本;在大致相同的交际情景中产生的具有相同信息性的双语文本;产生于不同语言文化环境,但属于相同的体裁和文本类型,并且具有相同功能的文本等等。简单地说,平行文本就是与原文内容接近的任何参考资料。

(2) 哪些语料库可供查阅科技文本?

现有许多公共科技英语语料库,如 WMT、IWSLT、LDC 等,可供译者查阅平行文本时使用。

2. 赏析下列翻译,指出译文的优点。

略。

3. 查找平行文本,翻译下列句子。

(1) The satellite will collect, grind and analyze around 70 samples of rock and soil from Mars.

(2) The Endoplasmic Reticulum (ER) is an organelle where proteins and lipids are synthesized and modified and serves as a calcium reservoir.

(3) Urban Rainwater Harvesting in those affected regions has dual benefits, supplementing municipal water supply and the potential to improve urban stream hydrology by capturing, consuming and effectively removing excess urban run-off.

(4) Graphics cards are notorious power hogs, and generate enough heat to warm a house.

(5) Nowadays, security requirements such as confidentiality, integrity, availability, authenticity, authorization and access control, non-repudiation, reliability, and privacy are also crucial in vehicular environments, given the latest wave of cyberattacks.

第三节　词　汇　翻　译

翻译下列句子，注意画线部分词汇翻译的方法。

(1) The versatility and robustness of ML algorithms enable them to adapt to virtually any application that has the basic requirement of an arbitrary dataset.

(2) Compared with non-NN algorithms, NN algorithms are highly efficient in terms of feature learning and extraction and require less manual input.

(3) Each neuron node is mathematically activated by the input stimulus data that are propagated throughout the NN.

(4) Chemicals such as formalin, hydrogen peroxide, and sodium hypochlorite are added to raw milk to increase shelf life.

(5) From Figure 3a-3, the transformed PLS components are linearly clustered together compared with MLR.

(6) It can be visualized on a PCA biplot on a 2D or 3D plane.

(7) The gas measurements from the unknown mixture are illustrated in red clusters.

(8) An RNN shares the same attributes as an FNN.

(9) The sensors in physical and chemical sensing systems require interaction with the target environment, which alters the chemical or physical properties of the sensors themselves to produce measurable signals.

(10) The 1D raw sensor data are stacked adjacently in rows to form a 2D signal image.

第四节　句　子　翻　译

翻译下列句子。

(1) In the past twenty years, we have witnessed the emergence of various revolutionary mobile devices, such as smartphones and wearable devices, which have led to the prosperity of mobile computing and allowed people to calculate and communicate anytime and anywhere.

(2) The increasing use of electronic media, as well as the decline in sleep quality and shortened sleep time, may play a core role in mental health. Although teenagers are facing changes resulting from urbanization and pressure from schools, these changes in their daily lives also have a certain impact on their mental health.

(3) The opportunities and complexity brought by the digital age may be unbearable by both industries and the market, for they are confronted by a large amount of potential information in every transaction.

(4) The overall goal of using intelligent learning methods is to train machines to think intelligently and make sensible decisions in both the same and different situations as humans.

(5) The rapid growth of data production in the digital era has introduced the concept of big data, which features remarkable capacity, diversity, accuracy, high speed and high value but also brings challenges to analysis. It is required to organize and deploy new analysis methods and tools to overcome the scale and complexity of different types of data.

第五节 修 改 译 文

1. 简答题。

(1) 修改译文的重点在哪些方面？

修改译文重点集中在词汇、句法、篇章等方面。

(2) 如何对译文进行审校？

译者在完成翻译任务后，首先需要对译文进行自我校对。首先对标点符号、单词拼写、语法等进行核查；其次检查专业术语前后是否一致，有无过译、漏译情况，句间逻辑是否清晰连贯；最后从语篇角度，检查译文整体风格是否与原文一致，语言是否通俗易懂、简洁明晰。自我校对后，需要寻求领域内专业人士或英语母语者在脱离原文的基础上对译文进行研读和润色，重点核查译文中信息是否翻译有误、英文表达是否流畅地道、符合规范。

2. 评析下列翻译。

略。

3. 翻译下列句子。

(1) It is actually a common convention in machine learning to split the data into segments.

(2) The most basic tensor is a one-dimensional tensor, which, in programming language, is called an array. It is an ordered series of individual numbers packed together.

(3) This is exactly what numerical stability requires in machine learning algorithms.

(4) Training a neural network usually requires a large amount of data, which is called a dataset. Datasets are generally divided into three categories: training set, validation set and test set.

(5) The main loss of an optical fiber system usually occurs in the electro-optical conversion of a transmitter and a receiver, and the actual cable loss is about one third of that of the coaxial cable.

第六节　译后编辑

1. 简答题。

(1) 译后编辑主要从哪些方面进行？

词汇、句子、语篇三方面。

(2) 常见的译后编辑需要修正的错误类型有哪些？（至少列出五个）

词汇：专业术语翻译错误；标点符号使用不规范；形式错误即文字或符号的表征形式不正确或不符合目标语言的表征习惯；词性判断错误等。

句子：长句不拆分，意思模糊；逻辑不连贯；信息冗余；句子译后比例失调等。

语篇：不符合目标语读者阅读习惯；文本风格不一致；过译和欠译等。

2. 翻译下列句子。

(1) There is a nucleus at the center of an atom. Around this nucleus revolve electrons.

(2) In the measurement of time intervals much shorter than a second, the decimal system is used.

(3) Overall, both studies showed that driving with the windows down has a significant negative effect on fuel efficiency—more than using the vehicle's air conditioner.

(4) The 4th United Nations World Data Forum opened in Hangzhou, the capital of east China's Zhejiang Province, on April 24. More than 1,000 representatives from over 100 countries and regions will attend the forum on-site, while online activities will attract more than 8,000 participants.

(5) The History Museum of China Aviation Industry was inaugurated in Beijing on April 16, becoming the only venue to provide a comprehensive display of China's 110-year aviation history.

第六章 科技翻译方法

第一节 逐词翻译法

1. 简答题。

略。

2. 翻译下列句子，尝试采用逐词翻译法。

(1) The software application requires a minimum of 4GB of RAM to run.

(2) The user interface has been redesigned with a new look and new feature.

(3) The keyboard has a built-in touchpad for navigation.

(4) The device has a built-in camera with zoom and autofocus capabilities.

(5) The software is compatible with both Mac and Windows operating systems, an advantage that expands its application fields.

(6) This artificial intelligence system can automatically learn and optimize to improve its performance and accuracy.

(7) This cloud computing platform can provide powerful computing and storage capabilities and can be used for large-scale data processing.

(8) This robot can autonomously walk and perform tasks without the need for human intervention, a situation that could only exist in our imagination.

(9) This database stores a large amount of information and can provide researchers with abundant data.

(10) This virtual reality system can give users an immersive experience and make them feel as if they are in different scenes.

第二节　直译翻译法

1. 简答题。

略。

2. 采用直译的翻译方法翻译下列句子。

(1) Artificial Intelligence (AI) technology includes fields such as machine learning, deep learning and natural language processing, which can be used to perform tasks such as recognizing images, speech, and text through algorithm training.

(2) Professional talents in the field of electronic information communication engineering need to have comprehensive knowledge and skills, and regard disciplinary integration and cross-domain application as development direction and goal.

(3) With the continuous improvement and application of artificial intelligence technology, the field of electronic information communication engineering will usher in a wider development space and application prospects.

(4) The wide application and development of electronic information technology has had a profound impact on industrial production, military security, public services and other fields.

(5) Challenges and problems such as information security risks continue to emerge, and they need to be solved through technical and management means to ensure the security and reliability of communication and data.

(6) The application of computer network technology in the field of electronic information communication engineering is very extensive, with great significance for the interconnection and data transmission of communication networks.

(7) Education and training in the field of computers are an important guarantee and foundation for the development of the field of electronic information communication engineering, requiring high-level faculty and advanced teaching facilities.

(8) The emergence and application of advanced technologies such as cloud computing and big data have brought unprecedented development opportunities and challenges to the field of the Internet of Things and shared virtual space.

(9) Global Positioning System (GPS) is a satellite navigation system that can determine the position and time on the ground through signals transmitted by satellites. GPS technology has been widely used in fields such as transportation, aviation, and military, improving the accuracy and reliability of location information and providing people with more convenient and safe travel experiences.

(10) Cloud computing is a computing method based on the Internet, which can achieve distributed computing and data storage. With cloud computing technology, users can quickly and elastically deploy and manage computing resources, improving resource utilization and flexibility. Currently, cloud computing has become an important technology used by many enterprises and individuals, promoting digital transformation and informationization construction.

第三节   意译翻译法

1. 简答题。

略。

2. 赏析下列翻译，指出译文的优点。

略。

3. 使用意译翻译法翻译下列句子。

(1) Recently, very fuzzy pictures of heavy atoms have been made with the use of a specially designed electron microscope.

(2) With its advantages of speed, efficiency and low cost, 3D printing technology is widely applied in the domains of medical treatment, industrial manufacturing, etc. In the medical field, it is increasingly used to build models of various human organs.

(3) The application of wireless transmission technology enables remote communication and data exchange between devices, playing a crucial role

in fields such as smart homes and the Internet of Things.

(4) The application of blockchain technology is particularly extensive in the financial industry due to its non-tamperable nature and decentralized advantages within its application scenarios.

(5) Through multiple iterations and optimized design, the algorithm has effectively reduced the need for manual intervention while ensuring efficiency and reliability, and to a certain extent, increased the accuracy of machine matching.

第七章 科技翻译工具

第一节 翻译工具概述

1. CAT 有哪些核心功能？

CAT 的核心功能包括翻译记忆、术语数据库、语言搜索、术语管理、交互式机器翻译、翻译对齐等。

2. 翻译记忆的基本原理是什么？

翻译记忆的基本原理是，译者在 CAT 翻译系统内进行翻译时，翻译记忆工具在后台工作，把完成翻译的内容储存在一个双语库中，译者可以在以后完整地使用这些内容或者根据需要进行调整。它首先将待翻译的文本划分为称为片段(segments)的单元，随着译者翻译文件的进展，软件将已经翻译过的片段存储到数据库中。当软件识别到待译片段与记忆库中已译的片段完全匹配或高度匹配时，就会建议译者复用或参考已译片段的译文，从而避免重复劳动。

第二节 语料库工具

1. 简答题。

(1) 语料库在翻译中有哪些主要用途？

语料库为译者提供了一个工作平台和参考工具，可用来提高译者的语言及文化意识。使用单语语料库，译者可以从更多的句子、语境中理解待翻译原文，并利用语料库统计分析结果中选择合适的译文。利用可比语料库核对有关翻译问题能够帮助译者提高表达能力，而且译文中的错误也会减少。其他作用还包括专业背景知识获取、术语的准确选用及目标语习惯表达等。

(2) 常见的科技语料来源有哪些？

常见的科技语料来源有中英文电子专著、教材、论文等，用于自行构建小型可比语料库和单语语料库。

2. 略。

3. 略。

第三节　计算机辅助翻译工具

1. 简答题。

(1) 主流的 CAT 工具通常包含哪些功能模块？

主流的 CAT 工具通常包含翻译记忆、术语管理、自动质量保证、机器翻译、译后编辑、项目管理等功能模块。

(2) CAT 的优点和缺点分别有哪些？

CAT 的优点：译文一致性高、翻译速度快、语言资产建立和维护简便、翻译总体成本低、支持文件类型丰富。

CAT 的缺点：割裂原文连续性，造成翻译中调整句序的困难；逐句段翻译习惯导致译员缺乏上下文观念，对机器译文和匹配译文过度依赖还会产生惰性；模糊匹配译文的可用性难以保证；应用领域有局限；软件成本和学习时间成本较高。

2. 略。

3. 略。

第四节　机器翻译引擎

1. 简答题。

略。

2. 利用两个以上机器翻译引擎翻译下列文字，对机器译文进行对比分析，然后搜索英文平行语料进行参照，对机器译文进行译后编辑，给出最终译文。

Hardness of Materials

The hardness of materials is measured in several ways, the simplest test for nonmetals being the scratch test. Substance A is harder than substance B if A will scratch B but B will not scratch A. A standard scale is used for representative minerals, with diamond the hardest, assigned the value 10 and talc, the softest, assigned the value 1.

diamond (C)	10	apatite($Ca_5(PO_4)_3F$)	5
corundum (Al_2O_3)	9	fluorite (CaF_2)	4
topaz ($Al_2SiO_4F_2$)	8	calcite($CaCO_3$)	3
quartz(SiO_2)	7	gypsum ($CaSO_4 \cdot 2H_2O$)	2
orthoclase($KAlSi_3O_8$)	6	talc($3MgO \cdot 4SiO_2 \cdot H_2O$)	1

There is great current interest in the development of materials of great hardness, for example as films for use as scratch-resistant coatings on lenses. It is widely felt that the scale between diamond and corundum is misleading, because diamond is much, much harder than corundum. It has been suggested that one might assign diamond the hardness 15, with the gap between 9 and 15 to be filled in eventually by synthetic materials, such as compounds of C and B.

Modern scales of hardness, such as the VHN scale, are based on indenter tests in which an indenter is pressed into the surface of the material and the size of the impression is measured.

第八章　科技翻译词汇处理

第一节　专 有 名 词

1. 简答题。

略。

2. 结合本节内容，翻译下列科技专有名词。

(1) Charpy impact test；

(2) Vickers hardness；

(3) 交流电；

(4) creep；

(5) elongation；

(6) 电子布告栏；

(7) algorithm；

(8) cloud computing;

(9) School of Communication Engineering;

(10) Chinese Institute of Electronics (CIE)。

第二节　准　确　性

1. 简答题。

略。

2. 结合本节内容，翻译下列句子，特别注意划线词翻译的准确性。

(1) If the quota issued is less than the actual emissions, enterprises will <u>purchase</u> quotas in the carbon trading market to meet their emissions demand.

(2) As the high-speed railway extends to the Midwest, the geological conditions are becoming more complicated, and the <u>engineering difficulties</u> keep increasing.

(3) Bank <u>client</u> confidentiality means that the banks have a duty to keep confidential all facts that involve their <u>clients</u>.

第三节　灵　活　性

1. 简答题。

略。

2. 结合本节内容，从词汇翻译灵活性的角度赏析下列句子的翻译。

提示：可从增译与省译、词性转换等方面考虑，言之有理即可。

第四节　多　样　性

1. 简答题。

略。

2. 结合本节内容，从词汇翻译多样性的角度分别给出下列句子的多种英译版本，尤其注意划线词的翻译。

(1) Now we are seeing <u>stunning progress</u>, with many countries reporting <u>dramatic</u> <u>reductions</u> in malaria case.

(2) The solution turned out to be advertising, and <u>it is not an exaggeration</u> to

say that Google is now essentially an advertising company, given that it is the source of nearly all its revenue.

(3) Terry Moffitt of Duke University and her research colleagues found that kids with self-control issues tended to grow up to become adults with <u>far more troubling set of issues to deal with</u>.

第五节　学　术　性

1. 简答题。

略。

2. 结合本节内容翻译下列句子，体现用词的"学术性"，尤其注意划线词的翻译。

(1) According to the U.S. Census Bureau, <u>approximately</u> 40 million people over 65 live alone.

(2) To this class of substance belong mica, porcelain, quartz, glass, wood, <u>etc</u>.

(3) However, most of the current research focuses on how to apply the affective factor in classroom activities, and there are <u>few studies</u> on how to reflect it in textbooks, <u>particularly</u> in science textbooks.

第九章　科技翻译句法处理

第一节　语　言　简　洁

1. 简答题。

略。

2. 赏析下列翻译，指出译文的优点。

提示：从使用名词化结构、用分词短语、独立主格结构或 with 结构来代替从句，使用介词结构以及使用代词的角度分析，言之有理即可。

3. 翻译下列句子，注意语言的简洁性。

(1) The application of information to the economy is best used in the networking of communication, or the so-called Net Economy.

(2) For example, the conversion of sound energy into electrical energy by a

telephone transmitter is a form of modulation.

(3) The violation of parity conservation would lead to an electric dipole moment for all systems.

第二节　表达清晰准确

1. 简答题。

略。

2. 赏析下列翻译，指出译文的优点。

略。

3. 翻译下列句子，注意使用清晰准确的表达。

(1) However, that risk is minimized because the time between updates is so short that it is unlikely that the output value being incorrect for such a short time duration will have a serious effect on process operation.

(2) According to Chart 1, in the past five years, great changes have taken place in people's diets. There is a general trend of decrease in the consumption of grain, while the consumption of some healthy food, such as milk, fruit and vegetable, is increasing steadily.

(3) Seventy percent profit was reported in this project.

第三节　统 一 性 原 则

1. 简答题。

略。

2. 指出并改正下列句子中的错误，优化句子使其更具统一性。

(1) Every subject has a defining benefit to us students. For example, math can improve our logical thinking and art can develop our cognitive and creative skills.

(2) People tend to live in a nuclear family but it can cause problems for the elderly.

(3) Highways and subways have been constructed, making it possible for people to travel from one place to another.

3. 判断下列句子是否遵循了统一性原则，如果不是，请对其进行修改。

(1) The silence of the forest was more oppressive than the heat. At this hour of the day there was no sound at all, not even the whine of insects.

(2) People with seasonal affective disorder can be treated by using light boxes where they are exposed to high doses of artificial light to mimic the brighter spring and summer mornings .

(3) It can be concluded that the Baud rate is very important to the telephone engineer as this rate establishes the type of telecommunication channel to be used.

第四节　长 短 适 中

1. 简答题。
略。
2. 赏析下列翻译，指出译文的优点。
略。
3. 翻译下列句子，注意灵活运用长短句式。

(1) The country is fighting inflation, a problem that is causing social and political tension.

(2) Not until the summer of 2018 did official government figures begin to acknowledge the scope of inflation.

(3) Today the vast majority of U.S. colleges and universities, public and private, prohibit guns.

第十章　科技翻译篇章处理

1. 简答题。
(1) 在对科技翻译篇章进行处理时，需要注意哪些问题？
科技翻译的篇章处理需要关注宏观意识、审美意识和程式化等问题。
(2) 在科技翻译语篇中，衔接和连贯分别指什么？
衔接指的是借助词汇或语法手段使文脉相通，形成篇章的有形网络，常见的衔接手段包括同义词重述、语意贯通、照应、逻辑连接词的使用等。连贯是指以信息发出者和接收者双方共同了解的情境为基础，通过逻辑推理来达到语义的连贯，形成篇章的无形网络。

2. 赏析下列翻译，指出译文在篇章方面的优点。

(1) 赏析：译文将原文中"想要确定"和"想知道"两个动作合二为一，将其后的三个宾语用 or 并列，这一灵活处理使译文更加简洁，译文中代词 it 或 its 与被指称对象精准照应，衔接紧密，最后一句用 for all these purposes 对前面列出的天体观测形式进行总结，承上启下，全文翻译时颇具宏观意识，译文语义连贯顺畅。

(2) 赏析：中文的三个例子在译文中分别转为 we、the rail network、a further example 作主语，每个句子的句式不是简单重复，而具有句式变幻的美感。中文中括号里的词在译文中也直接转为同位语翻译，处理巧妙。

(3) 赏析：译文使用 since、if、however、since、but 等连接词或副词贯穿全文的逻辑关系，其中穿插定语从句、介词短语、比较结构等多种表达，表意灵活，手段丰富，语义精准呈现，衔接完美，尽显翻译的宏观意识和整体美感。

3. 请赏析节选自 2023 政府工作报告的中英双语版，总结译文的程式化特点。

本文的程式化体现在：原文中的汉语无主句在译文中多采用第一人称复数作为主语；多用并列的词或短语；用词规范精准、表达较为固定；翻译有时采用省译、合译等方法；多用长句；时态以一般过去时为主；多用 need、should 等情态动词。

第十一章　翻译腔应对策略

第一节　直译与意译相结合

1. 简答题。

(1) 产生翻译腔的原因有哪些？

科技翻译中的翻译腔可能由以下原因产生：① 译者英语熟练程度不高；② 译者不够熟悉汉英两种语言之间的差异及背后的中西思维模式差异；③ 译者忽略中西方文化差异；④ 译者对英语写作原则、翻译技巧等知识不够了解或理解有偏差；⑤ 译者崇尚翻译腔；⑥ 译者对源语文本理解不够准确；⑦ 译者缺乏翻译实践经验。

(2) 直译的优点和缺点分别有哪些？

优点：能够传达原文意义，体现原文风格等。

缺点：导致译文不通顺或不符合目标语表达习惯，影响信息传递效果。

2. 赏析下列翻译，指出译文的优点。

略。

3. 翻译下列句子，注意避免生硬直译。

(1) With cities getting smarter, there is an increasing demand for surveillance and security. Accordingly, cross-modality person re-identification, one of the key techniques in this field, has gained considerable attention from the academic circle and industrial sector in recent years.

(2) All-flash storage features better manageability and maintainability. In addition, SSDs deliver more flexibility in size and are available at multiple lengths, widths, and heights.

(3) Facing new service requirements, the memory-driven infrastructure is going mainstream.

(4) Modern query optimizers require detailed statistics of the structure and make of data in tables to enable them to make the "optimal" decision on how to execute complex queries.

(5) IT infrastructures are prone to failures such as server faults, disk crashes or storage corruption sites, outages and human error that can incur costly unplanned downtime.

第二节　形合与意合相结合

1. 赏析下列翻译，指出汉语的意合如何转换为英语的形合。

略。

2. 翻译下列句子。

(1) This technology has gradually matured, laying a solid foundation for its large-scale implementation.

(2) Quantum mechanics is a branch of physics that describes the strange behavior of matter and energy on the tiniest scales.

(3) The basic concept of the 3CS system is to divide security control domains based on the processes of each cloud service module, enabling security

control requirements to be embedded into the cloud service management process, which in turn ensures that security management responsibilities are clear, measurable, and traceable.

(4) Based on the shared responsibility model, Huawei Cloud continues to build and enhance its security compliance capabilities in its infrastructure (across the physical environment, network, and platform layers) and cloud services to ensure the security and compliance of its services and data.

(5) This makes the memory space a large attack surface open to attackers who could successfully compromise any of the software modules by exploiting a memory corruption vulnerability in real-time microcontrollers.

(6) The seventh China Grand Awards for Industry were announced on March 19, with 19 Chinese enterprises and 19 projects named as winners.

(7) 3D printing or additive manufacturing is a family of technologies that employ a virtual Computer Aided Design (CAD) model to create a physical object through the consecutive creation of layers.

(8) China Telecom has been building an all-optical network, with gigabit optical networks now covering more than 300 cities.

(9) The development of China's large-scale AI models is booming, with several technical routes making breakthroughs at the same time.

(10) The sharing, circulation, trading and application of data, which is regarded as a new type of production factor, are key to promoting the country's digital development.

第三节　动词与名词化结构相结合

1. 赏析下列翻译，指出动词和名词化结构如何相结合。

略。

2. 翻译下列句子，注意不要过度使用名词化结构。

(1) We can now define a causal effect for an individual.

(2) We have successfully implanted an interferometric pressure sensor in a New Zealand white rabbit and directly measured the intraocular pressure of one of the rabbit's eyes.

(3) To further investigate this discrepancy, we employed a second method to measure the residual stress in the SiN thin film using resonant frequency and displacement measurements.

(4) It is not possible to process traffic at these throughputs using CPU cores due to the number of compute resources required.

(5) Among various types of embedded systems based on microcontrollers, real-time microcontroller systems are designed to meet the deadline constraints of real-time processes.

第四节 主动语态与被动语态相结合

1. 阅读下文，分析具体语境中使用主动语态或被动语态的原因。

略。

2. 赏析下列翻译，指出译文的优点。

略。

3. 以适当语态翻译下列句子。

(1) The specific contributions of this paper are as follows. We trained one of the largest convolutional neural networks to date. We wrote a highly optimized GPU implementation of 2D convolution and all the other operations inherent in training convolutional neural networks, which we make available publicly.

(2) The highly challenging task of estimating a high-resolution (HR) image from its low-resolution (LR) counterpart is referred to as super-resolution (SR). SR received substantial attention from within the computer vision research community and has a wide range of applications.

(3) Different from previous works, we define a novel perceptual loss. An example photo-realistic image that was super-resolved with a 4× upscaling factor is shown in Figure 1.

(4) Dong et al. used bicubic interpolation to upscale an input image and trained a three-layer fully convolutional network end-to-end to achieve state-of-the-art SR performance. Subsequently, it was shown that enabling the network to learn the upscaling filters directly can further increase its performance both in terms of accuracy and speed.

(5) The architecture of our network is summarized in Figure 2. It contains eight learned layers—five convolutional and three fully connected. Below, we describe some of the novel or unusual features of our network's architecture.

第十二章　常见问题分析

第一节　科技知识欠缺

1. 简答题。

略。

2. 翻译下列句子，注意运用专业知识。

(1) However, the satellite attitude system is affected by many uncertain factors such as environmental stochastic disturbance torque, uncertainty of the system inertia matrix, and disturbance torque caused by flexible deformation.

(2) The 125 MW high flux engineering test reactor and the 60 MW China Advanced Research Reactor (CARR) have produced a wide range of medical and pharmaceutical radioisotopes, and boosted the development and application of neutron imaging.

(3) The key width, result width, and number of entries in a table decide the amount of on-chip logic resources and memory (SRAM/DRAM) used on the FPGA.

(4) With the evolution of policy-based networks and the emergence of software-defined networks (SDN) and intent-based networks (IBN), the firewalls of different throughputs and features are being deployed at many different locations within the enterprise network.

(5) Traffic originating from multiple applications, accessing the different nodes of enterprise and data center networks, needs to be classified based on the type of traffic.

第二节　英语能力不足

1. 找出并修改下列学术论文中的错误。

Micro-Expression Recognition Based on the MAML Meta-Learning Algorithm

The micro-expression is an accurate and effective behavioral clue that can reveal true changes in people's emotion. It is widely used in education, interrogation, clinical diagnosis etc. Previous researches on micro-expression recognition focused on traditional image recognition and deep-learning methods, both of which rely on large amounts of training samples and have lower accuracy. Therefore, we introduce a meta-learning training method to achieve high accuracy with small samples. A three-dimensional convolutional neural network is used as the representation of the learning input of the feature extractor. In the meta-training stage, the optimal parameter$\theta*$ is obtained through double-loop iterative updating. We have conducted extensive experiments on two typical micro-expression datasets: CASME II and SMIC-HS. The results show recognition accuracy based on meta-learning can reach 92.97%, which is significantly better than that of traditional algorithms. In addition, the model can adapt to new micro-expression recognition tasks flexibly and quickly with small samples.

The research on micro-expressions had not started until lately, and early researches mostly used traditional machine learning methods, mainly to obtain subtle facial motion information by hand. Although this method has some validity in recent researches, it consumes a lot of manpower and relies too much on researchers' prior experience. In recent years, with the rapid development of computer vision, researchers have applied deep learning and other related technologies to micro-expression recognition, thereby improving the recognition accuracy. At present, the collection of micro-expression sample data is faced with such problems as short duration, small motion range, and large impact by environmental factors, which makes it difficult to collect data with a small sample size and to conduct better deep learning training. Therefore, how to extract the effective features of micro-expressions from tiny facial movements has become researchers' focus. In addition, the collection of micro-expression data requires videos with specific emotions to induce the subjects to express and maintain specific types of expressions. Professionals analyze and observe the time when the micro-expressions are generated and the subjects' true emotions to obtain sample data. The micro-expression data sampling process is complicated and cumbersome and demanding on the environment, which causes great difficulties to the large-scale sampling of data sets. Therefore, the massive collection of micro-expression data is also one of the problems to be solved. The last problem to be solved in the study of micro-expression recognition using deep learning methods is how to use the powerful feature extraction capabilities of neural networks while alleviating the network's dependence on the amount of data.

2. 翻译下列段落，注意语言的准确性、丰富性、连贯性、得体性。

(1) The microwave apparatus used in the radar system is characterized by its complicated structure, high precision and high cost. This paper introduces the precision manufacturing technology of a complicated microwave apparatus, presents and discusses the key points of the manufacturing process with emphasis on the major reasons for machining deformation and the measures taken in deformation control. This manufacturing technology has proved to be successful in practice, thus offering valuable reference in resolving similar problems.

(2) Secure aggregation is a cryptographic protocol that securely computes the aggregation of its inputs. It is pivotal in keeping model updates private in federated learning. In this work, we show that a malicious server can easily elude secure aggregation as if the latter were not in place. We devise two different attacks capable of inferring information on individual private training datasets. The attacks are generic and equally effective regardless of the secure aggregation protocol used. Our work demonstrates that current implementations of federated learning with secure aggregation offer only a "false sense of security".

(3) We introduce Co-Fusion, a dense SLAM system that takes live stream of RGB-D images as input and segments the scene into different objects (using either motion or semantic cues) while simultaneously tracking and reconstructing their 3D shape in real time. We use a multiple model fitting approach where each object can move independently from the background and still be effectively tracked and its shape fused using the information

from pixels associated with that object label. Previous attempts to deal with dynamic scenes have typically considered moving regions as outliers, and consequently do not model their shape or track their motion over time. In contrast, we enable the robot to maintain 3D models for each of the segmented objects and to improve them over time through fusion.

(4) Huawei, the world's leading telecom service provider, has worked tirelessly to maintain robust network operation and uninterrupted service for its users. Huawei has relied on its AI capabilities and digital platforms to ensure stellar digital delivery and maintenance to help carriers better engage with their customers through improved service quality. Its AI customer service, online healthcare capabilities, and new marketing approaches have made a difference for carriers. This year, Huawei has deployed hundreds of thousands of 5G and 4G sites worldwide, all based on its Integrated Service Delivery Platform (ISDP). So far, over 100 apps and 500 microservices are available on ISDP; these cover all the delivery phases, including PO acquisition, survey, delivery, acceptance, invoicing, and payment collection, minimizing close exposure and the public health risk that comes with it.

(5) Information drove the rise of radio and television. In the 1880s, Heinrich Hertz demonstrated that radio waves are a form of electromagnetic radiation (as predicted by Maxwell's theory). In the 1890s, Indian physicist Jagadish Chandra Bose conducted an experiment using microwaves to ignite gunpowder and ring a bell, proving that electromagnetic radiation could travel wirelessly. These scientific insights laid the foundation for modern telecommunications.

A similar evolution occurred in computing technology. Early mainframe computers were so large that they took up most of the space in a room. They were expensive and did not dissipate heat well, requiring cooling. They seemed to be affordable only to governments or large corporations with deep pockets. In the 1980s, personal computers dramatically changed this scenario. Suddenly, any business and many people could buy and use computers, not only for intensive computing but also for information management.

第三节　母语负迁移

1. 赏析下列翻译，指出译文的优点。

略。

2. 翻译下列句子，注意避免母语负迁移。

(1) Renowned quantum physicist Pan Jianwei had been assembling an academic team at USTC since 2001, and in 2009, he proposed that the university help found a company to continue the group's research and commercialize its findings.

(2) Quantum computers can take computational shortcuts when simulating extremely complex scenarios whereas conventional computers have to find a solution step by step, taking significantly more time in the process.

(3) Construction of virtual power plants is on the rise in China as the country continues to move toward a more sustainable energy mix, with renewables taking up an increasing share.

(4) China's computing power sector is expected to witness speedy growth as tech companies are scrambling to roll out ChatGPT-like artificial intelligence chatbots, which necessitates higher requirements for computing capacity in the process of large language model training and operation.

(5) However, as the focus of power generation gradually shifted toward renewable energy resources, such as solar and wind, problems emerged as these modalities are intermittent and not always available when needed, making it extremely difficult to manage power supply.

第四节　职业精神不够

1. 赏析下列翻译，指出译者精神的体现。

略。

2. 指出下列翻译中有失译者职业精神的体现。

(1) "模件"并非"模型"，应译为 module 而非 model。译者对专业术语未能仔细斟酌及查证。

(2) "CO2" 应为 "CO_2"。译者未关注细节。

(3) "蓬勃发展" 为典型汉语四字词,但逐词翻译不够简洁。可译为 flourish 或 develop vigorously。可见译者未能主动避免母语负迁移。

(4) 译者逐词翻译,未能调整语序,突出重点,可见未能发挥译者能动性。这句话可译为：China's newly installed combined wind and solar power capacity reached a record 125 million kilowatts in 2022, bringing the tally of total installed capacity to over 1.2 billion kW, as the country stepped up efforts to ensure energy security while facilitating green energy transition.

(5) 原文 "提供" 显然为手误,应为 "提高",译为 improves。可见译者未能妥善处理原文的错误信息。

(6) 术语翻译错误,译者缺乏查证意识和专业精神。这句话可译为：It can be seen from this curve that the law of product failure can be divided into three stages: early failure period, intrinsic failure period and wearout failure period.

(7) 该译文采用直译法,语序与汉语语序相同。两个状语堆积在句首,主语部分过长,句子焦点不够清晰。受汉语标点符号影响,论文题目放入引号内,整句话过长,可读性不强。译者缺乏译者主动性。这句话可改为：Recently, Professors Zhang Jing and Han Ming and their respective teams co-authored a review paper Deep Learning for Visible-Infrared Cross-modality Person Re-Identification: A comprehensive Review in top journal *Information Fusion* (IF:17.564) at the invitation of Professor Salvador Garcia, Editor in Chief of *Information Fusion*. Professor Zhang's team comes from the School of Mechano-Electronic Engineering, *** University, and tenured professor Han's team, from the Computer Science Department, Aberystwyth University.

(8) 译文完全按照汉语语序,导致第一句话两个分句主语不同,行文不够简洁流畅；第二句话和第一句话逻辑关系不清晰；第二句话 while 之前的部分过长,导致句子头重脚轻；修饰动词 support 的状语 in a reliable way 离动词太远可能造成读者困惑；第一句和第二句重复出现 support。译者缺乏译者主动性。这句话可改为：Relied upon by some of the world's largest repositories, our platform turns storage challenges into business advantages. It does this by reducing storage costs while reliably

supporting both traditional and emerging cloud-born workloads for enterprise mobile, social, analytic and cognitive computing.

3. 以译者的专业精神翻译下文。

Apple has unveiled digital wellbeing tools to help people reduce the time they spend glued to their screens. A new app called Screen Time will offer iPhone and iPad users a dashboard highlighting how much time they have spent using which apps, how many notifications they receive, how often they pick up their device and how their usage patterns compare to the average. The app also lets users set daily time limits for individual apps, and a notification will be shown when the time limit is about to expire. Parents will be able to access their children's activity reports from their own devices to understand and manage their browsing habits. Screen Time will be available with iOS 12, the latest version of Apple's mobile operating system, which will be launched later this year.

The company also doubled down on its commitment to privacy with the introduction of tools to prevent advertisers from tracking users of Apple devices from being tracked online. The company is also clamping down on a technique called "fingerprinting" which allows advertisers to identify and track individual web users based on the unique way their computer is set up, including the device, operating system and fonts they are using.

参 考 文 献

[1]　American Translators Association. Translator competence[2017-06-13]. https://www.atanet.org/translation/translator-competence.

[2]　ASTON G. Corpus use and learning to translate[J]. Textus, 1999(12):1000-1025.

[3]　BERMAN A. The experience of the foreign: culture and translation in romantic Germany[M]. New York: State University of New York Press, 1984/1992.

[4]　BERMAN A. Translation ethics// GAMBIER Y, DHULST L. History of translation knowledge: sources, concepts, effects[M]. Amsterdam/Philadelphia: John Benjamins Publishing Company, 2018.

[5]　CHRISTINA S, BEVERLY A. Developing translation competence: introduction [M]. Amsterdam/ Philadelphia: John Benjamins Publishing Company, 2000.

[6]　HALLIDAY M A K, HASAN R. Cohesion in English (1st ed.)[M]. London &New York: Routledge, 1976.

[7]　KIRALY D C. Pathways to translation: pedagogy and process[M]. Kent State University Press, 1995.

[8]　LYNNE B. Towards a methodology for a corpus-based approach to translation evaluation[M]. Meta: Journal des traducteurs, 2001.

[9]　NIDA E A. Toward a science of translating[M]. Shanghai: Shanghai Foreign Language Education Press, 2004.

[10]　NIDA E A, TABER C R. The theory and practice of translation[M]. Shanghai: Shanghai Foreign Language Education Press, 2004.

[11]　NORD C. Translating as a purposeful activity: functionalist approaches explained[M]. Shanghai: Shanghai Foreign Language Education Press, 2018.

[12]　PYM A. On translator ethics: principles for mediation between cultures[M]. Amsterdam/ Philadelphia: John Benjamins Publishing Company, 1997/2012.

[13]　BELL R T. Translation and translating: theory and practice[M]. Beijing: Foreign Language Teaching and Research Press, 2001.

[14]　SELINKER L. Interlanguage[J]. International review of applied linguistics in language teaching, 1972(10): 209-232.

[15]　HORNBY M S. Translation studies: an integrated approach[M]. Shanghai:

Shanghai Foreign Language Education Press, 2001.

[16] VENUTI L. The scandals of translation: towards an ethics of difference[M]. London: Routledge, 2002.

[17] 陈坚林. 计算机网络与外语课程的整合：一项基于大学英语教学改革的研究[M]. 上海：上海外语教育出版社，2010.

[18] 陈志杰. 翻译伦理学研究[M]. 北京：科学出版社，2021.

[19] 冯志杰. 汉英科技翻译指要[M]. 北京：中国对外翻译出版公司. 1999.

[20] 付吟璐. 专有名词汉译规范化问题的探讨[D]. 天津：天津大学，2012.

[21] 傅勇林，唐跃勤. 科技翻译[M]. 北京：外语教学与研究出版社，2012.

[22] 黄友义，杨平，邢玉堂. 创新翻译专业教育模式，培养新时代高层次翻译人才：翻译专业学位博士研究生培养调研座谈会综述[J]. 中国翻译，2023，44(1): 12-13.

[23] 莱斯. 翻译批评：潜力与制约[M]. 上海：上海外语教育出版社，2004.

[24] 冷冰冰，王华树，梁爱林. 高校 MTI 术语课程构建[J]. 中国翻译，2013，34(1): 55-59.

[25] 李长栓. 非文学翻译理论与实践[M]. 北京：中国对外翻译出版有限公司，2012.

[26] 连淑能. 中西思维方式：悟性与理性：兼论汉英语言常用的表达方式[J]. 外语与外语教学，2006，7: 35-38.

[27] 刘宓庆. 文体与翻译[M]. 北京：中国对外翻译出版有限公司，2012.

[28] 刘宓庆. 文化翻译论纲[M]. 武汉：湖北教育出版社，1999.

[29] 陆谷孙. 英汉大词典[M]. 2 版. 上海：上海译文出版社，2007.

[30] 罗新璋，陈应年. 翻译论集[M]. 北京：商务印书馆，2009.

[31] 马会娟. 汉译英翻译能力研究[M]. 北京：北京师范大学出版社，2013.

[32] 穆雷. 也论翻译研究之用[J]. 中国翻译，2012，33(2): 5-11.

[33] 纽马克. 翻译教程[M]. 上海：上海外语教育出版社，2001.

[34] 纽马克. 翻译问题探讨[M]. 上海：上海外语教育出版社，2001.

[35] 诺德. 目的性行为：析功能翻译理论[M]. 上海：上海外语教育出版社，2001.

[36] 潘鸣威. 英语书面表达能力特征量表研究[J]. 外语界，2022(6) : 14-16.

[37] 彭萍. 翻译伦理学[M]. 北京：中央编译出版社，2013.

[38] 郤春生. 汉英科技翻译如何体现客观性要求[J]. 中国科技翻译，2003，16(4): 15-17.

[39] 任朝迎. 功能句子观视角下的汉英科技翻译[J]. 上海翻译，2017(1): 36-40.

[40] 任文. 新时代语境下翻译伦理再思[J]. 山东外语教学，2020(3): 12-13.

[41] 任文. 机器翻译伦理的挑战与导向[J]. 上海翻译，2019(5): 48-50.

[42] 沙特尔沃思，考伊. 翻译研究词典[M]. 谭载喜，译. 北京：外语教学与研究出版社，2005.

[43] 单其昌. 汉英翻译技巧[M]. 北京：外语教学与研究出版社，1990.

[44] 泰特勒. Essay on the principles of translation[M]. 北京：外语教学与研究出版社，2007.

[45] 王少爽. 翻译专业学生术语能力培养：经验、现状与建议[J]. 外语界，2013(5): 30-34.

[46] 韦孟芬. 英语科技术语的词汇特征及翻译[J]. 中国科技翻译，2014，27(1): 5-7.

[47] 严俊仁. 汉英科技翻译新说[M]. 北京：国防工业出版社，2010.

[48] 杨荣广. "翻译伦理"概念的批评与反思[J]. 解放军外国语学院学报，2022，45(2): 114-115.

[49] 译员职业道德准则与行为规范[M]. 北京：中国标准出版社，2019.

[50] 于建平. 科技论文汉译英中若干问题分析[J]. 中国翻译，2001(1): 32-34.

[51] 袁晖，李熙宗. 汉语语体概论[M]. 北京：商务印书馆，2005.

[52] 赵稀方. 二十世纪中国翻译文学史. 新时期卷[M]. 天津：百花文艺出版社，2009.

[53] 中国翻译工作者协会，《翻译通讯》编辑部编. 翻译研究论文集(1894—1948) [M]. 北京：外语教学与研究出版社，1984.

[54] 中国日报. 英语点津. https://language.chinadaily.com.cn/.